THE SOCIAL IMPACT OF THE
CHERNOBYL DISASTER

The Social Impact of the Chernobyl Disaster

David R. Marples

Adjunct Professor of Slavic and
East European Studies
and Research Associate
Canadian Institute of Ukrainian Studies
University of Alberta

Introduction by Victor G. Snell

MACMILLAN

First published 1988
Reprinted 1990

Published by
MACMILLAN ACADEMIC AND PROFESSIONAL LTD
Houndmills, Basingstoke, Hampshire RG21 2XS
and London
Companies and representatives
throughout the world

Printed and bound in Great Britain by
Antony Rowe Ltd
Chippenham, Wiltshire

British Library Cataloguing in Publication Data
Marples, David R.
The social impact of the Chernobyl disaster.
1. Ukraine. Chernobyl. Nuclear power
stations. Accidents, 1986
I. Title
363.1'79
ISBN 0–333–46421–4 (hardcover)
ISBN 0–333–48198–4 (paperback)

TO MY PARENTS,
Ella and Neville Marples

CONTENTS

PREFACE

The Canadian author Robertson Davies once remarked that he was moved to write a certain story because it was like a little black poodle that would not go away. He would think about that poodle daily and dream about it at night, until finally he felt obliged to commit his thoughts to paper in the form of a book. The Chernobyl disaster has been my own "black poodle." The ramifications of that accident have been immense and far-reaching. When I published *Chernobyl and Nuclear Power in the USSR* in late 1986, new information about the accident was still emerging. After a further six months had transpired, I realized that I had considerably more information about the aftermath of the accident than about the events leading up to it (the main subject matter of the first book). Heedless of a colleague's advice that it does one's career no good to write two books on the same subject, I decided to write this second monograph.

I had other reasons for this second study beyond the information gathered from various sources. The key one was the interpretation of the accident both in the Soviet Union and the West. Throughout 1986 and 1987, I gave talks and attended meetings about Chernobyl, from the U.S. State Department and Atomic Energy of Canada Limited to the American Political Science Association and even to small Ukrainian communities, such as that in Sudbury, Ontario. I have spoken with the most learned of scientists at the U.S. Nuclear Regulatory Commission and the most earnest of students at universities such as Michigan and Pennsylvania and even schoolchildren at a Ukrainian bilingual school in Warren, Michigan. In every case, I felt there were gaps in the information about the disaster. I could not claim in any respect to have all the answers to the questions. Rather my intention was to answer those questions that fall into my area of expertise.

Chernobyl was a media event. Although I am not a journalist, I could not fail to be aware of the peculiarities and quantum leaps in the interpretations of the accident in the Western media. There was the initial horror, followed by widespread praise for the relative openness of Soviet accounts, culminating in the August 1986 IAEA meeting in Vienna, which was a public relations triumph for the Soviets that exceeded the December 1987 summit between Mikhail Gorbachev and Ronald Reagan. At the latter, the Soviets had first been obliged to make substantial concessions from their original bargaining position. In Vienna, I felt, it was almost forgotten why the accident occurred in the first place, which was the USSR's overdependence on the development of atomic energy at all costs, the only explanation for unqualified operators being on the staff of a major nuclear power plant.

In October and November 1987, I joined various American media representatives in a tour of the Soviet Union that involved intensive discussions with Soviet officials on a variety of questions. I attended a press conference of the Soviet Academy of Sciences and spoke with numerous representatives of the Soviet media at both formal and informal gatherings. My impression was that in many respects the Soviets themselves, below what might be termed the "scientific elite," had many questions about the Chernobyl disaster and were far from satisfied by the answers received thus far. The visit made me even more determined to produce the current book.

In Moscow in particular, I noted two distinct currents of thought that crossed the usual dividing lines perceived between the conservatives and the supporters of glasnost. There was the "traditional" wave, which consisted of those people who supported glasnost because it was the current policy and espoused personally by General Secretary, Mikhail Gorbachev. And there was a second current, which genuinely sought openness on all questions, no matter what the outcome. Boris Yeltsin is the obvious example of the latter current, but on the issue of Chernobyl, there are others, less prominent, but no less determined. One is Andrei Pralnikov of *Moscow News*, and another is Stanislav Kondrashov of *Izvestiya*. Kondrashov, with whom I was able to talk during one of our group sessions, was by his own account totally dissatisfied with the Soviets' decision not to reveal information about the Chernobyl disaster in the first days after it occurred. He was in Poland at the time, and thus well aware of the fallout and its consequences.

It is thanks to Soviet reporters such as Kondrashov, Pralnikov, Vladimir Gubarev and Ukrainian writers such as Yurii Shcherbak and Oles Honchar that we have gleaned a more accurate picture of what happened after Chernobyl. On the other hand, these writers are still restricted in what they can report. My own aim was to try to cross this invisible information barrier, and to use the Soviet sources of various hues and quality to put together a more complete picture of what has happened since April 26, 1986.

It should be added that Chernobyl was an event of such complexity and so many dimensions that it is not possible for any one scholar to be an expert on every dimension. I confess to my ignorance in the fields of medicine and reactor technology. As far as the latter is concerned, I have relied on the services of Dr. Victor Snell, the Manager of Safety Engineering of the CANDU-300 Project at Atomic Energy of Canada Limited (AECL), who was a member of the Canadian delegation to the IAEA meeting in Vienna in August 1986. Dr. Snell provides an expert AECL version of how the disaster occurred in the Introduction to this book. It should be emphasized that although the Introduction represents AECL's opinions on the accident, the content and views expressed in the remainder of the book are entirely my own.

On medical affairs, which are included in Chapter 1 of this book, I have simply examined the various opinions expressed without contributing my own, unless to comment on some odd statement or obvious discrepancy. It seemed essential to include such a chapter, if only to demonstrate the different views on the ultimate casualty figure from Chernobyl, a matter of the utmost controversy.

The remaining chapters try, above all, to show what happened to the victims of the Chernobyl disaster, how people's lives were changed. Chapter 2 looks at the environmental impact. Chapter 3 examines the economic and political consequences of Chernobyl and how it affected the Soviet nuclear energy program. Chapter 4 provides an account of the portrayal of the disaster in the Soviet Union on film and on paper, and describes the lives of the evacuees, and the heroes and villains of Chernobyl.

In Chapter 5, we analyze one of the most controversial and little covered aspects, namely work in the special zone. The chapter includes a section on the "Estonian affair" and the question whether the rights of the workers were violated. Chapter 6 concerns reconstruction in the zone and the restarting of the Chernobyl plant, concluding with a detailed examination of the building of the new city for plant operatives, Slavutych. The final chapter looks at the changes made to the RBMK reactor, whether operator training has improved and the emerging opposition to nuclear power in the Soviet Union.

Again, I have been heavily reliant on Soviet sources. Interestingly, a reviewer for the British journal *Nature* felt that the first book was too critical of the Soviets and reflected a certain bias on my part. While I am not averse to constructive criticism, this comment seemed to me somewhat far-fetched in that all the comments on poor workmanship at Soviet nuclear power plants came directly from Soviet sources. They were not invented by me. I dare say that if I was a student of the North American nuclear industry rather than of Soviet energy questions, I would be equally as critical if all the sources stated the same thing, namely that nuclear stations were being built without due regard for the safety of personnel.

Moreover, it should be stated that there have been comments made by Soviet scientists that disturb me because they appear too sweeping and too concerned to assuage the anxieties of the people of the Chernobyl region. Romanenko's comment, for example, that there are no health concerns in the area as a result of the accident can hardly be taken seriously; nor can Leonid Ilyin's statement that Prypyat was not evacuated for 40 hours because the radiation situation did not present a danger to the population. If the writer did not feel that certain Soviet accounts of the accident's aftermath were misleading, then this book would not have been written.

It has frequently been stated that Chernobyl provided a warning to the world of what would happen in the event of a nuclear war. At the risk of playing the devil's advocate, permit me to say that it did no such thing. The fallout from the accident can in no way be compared to the unleashing of nuclear missiles. Chernobyl was a major disaster and one of the few twentieth-century disasters on a world-scale. But it provided a lesson for the nuclear power industry rather than about nuclear weapons. It revealed the fatal consequences of running nuclear power plants improperly and of an insufficiently vigilant attitude toward the dangers of the atom.

Therefore the lesson to be derived from Chernobyl is that safety should be uppermost in the minds of those who run the nuclear power industry in various countries of the world. It is hardly my task to comment on the pros or cons of nuclear energy, or to state whether it can be made safe. The question I have tried to answer is whether Soviet thinking on the industry has changed fundamentally as a result of the disaster. In the final analysis, that could be the only possible positive outcome of Chernobyl.

NOTE ON TRANSLITERATION

For the most part, this book transliterates Ukrainian place names and personalities according to their Ukrainian spelling. An exception has been made for those names that are familiar in English only in their Russian form, such as Chernobyl, the Dnieper River, and Kiev. In block quotes from Ukrainian sources, however, the original Ukrainian form is retained: hence Chernobyl becomes Chornobyl and the Dnieper becomes the Dnipro River.

ACKNOWLEDGEMENTS

This book was written with the financial backing of the Ukrainian National Association, Jersey City, New Jersey, a nonpolitical fraternal organization with membership across the United States and Canada. I am especially grateful to the organization's Executive Committee for its generous and unqualified support.

Other individuals and organizations have assisted in various ways, and in particular I would like to mention: Paul Goble, *National Geographic*, the *San Francisco Examiner*, Arnold Beichman, Dzintra Bungs, Toomas Ilves, Keith Bush, Jurij Dobczansky, Andrew Hruskewycz, Bohdan Krawchenko, John Fox, Lucy Kerner, the Sygma Agency, Harold Denton, Miller B. Spangler, Richard Wilson, and the editorial board of *Izvestiya* newspaper in Moscow.

The staff at Atomic Energy of Canada Limited (AECL), Mississauga, Ontario, provided me with a different perspective of the accident during a conference in April 1987, and permitted Victor Snell to write the opening chapter of this book. I am indebted to Dr. Snell for devoting time and energy to his task and for providing a chapter of perceptive originality on an extremely complex subject. While AECL may not necessarily agree with the interpretations expressed in the book itself, I hope that our association may long continue.

I have been encouraged especially by the interest and concern of students of Ukrainian background about the events discussed in this book. I am indebted to Ksenia Kozak, President of the Ukrainian Students' Association of the University of Michigan, for organizing a lecture tour that enabled me to talk to so many of these students in Ann Arbor and on the eastern seaboard of the United States. Their interest fueled my own and it is for them that this book was written.

A special thanks goes to my assistant during the period of research on this book, Leda Hewka, a graduate of the University of Pennsylvania who was often asked to work long hours over the summer of 1987, and did so willingly, and with great ability; and to Roman Solchanyk of Radio Liberty, who again supplied me with an enormous quantity of source materials on Chernobyl and related topics.

Finally, my wife Lan has had to suffer my preoccupation with Chernobyl for two years now, but has done so with patience and understanding.

INTRODUCTION

The Cause of the Chernobyl Accident

VICTOR G. SNELL

The Perspective

On August 25, 1986, hundreds of nuclear scientists and engineers converged on the modernistic offices of the International Atomic Energy Agency (IAEA) in Vienna, Austria. We (I was fortunate enough to be among them) were so numerous that we overflowed the largest meeting room the IAEA had to offer, and thus were housed in two huge rooms with closed-circuit TV to help us communicate. Why was there so much excitement? For the first time since the Chernobyl nuclear power station accident of April 26, 1986, we were to find out *from the Soviets themselves* what had really happened.

The answers went beyond what most of us had ever guessed.[1] Yet tantalizing holes still remained in the story. Now, over a year later, thanks to intensive work in Canada, the United States and other Western countries, as well as in the Soviet Union, many of those holes have been filled in, and we believe we know in detail what went wrong. The Canadian work in discovering the most likely root cause of the accident has now been accepted by most of the Western world and acknowledged by the Soviet Union.

First, however, we will go back into the nature of the Soviet nuclear program, look critically and fairly at the design of the Chernobyl nuclear plant, and describe the sequence of events on the night of April 25-26. Like all accidents, it resulted from a combination of human error and

1

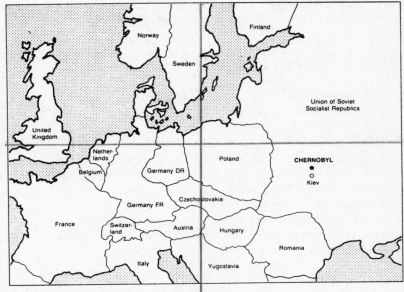

Figure 1(a) Chernobyl reactor location.

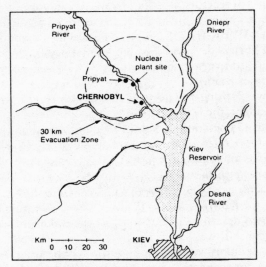

Figure 1(b) Area nearby the Chernobyl reactor site.

design weaknesses; like all accidents, it could have been stopped at a number of places and the world would never have heard of it.

Chernobyl—The Place and the Plant

The Chernobyl Site

Chernobyl itself is a small town of 12,500 people in the Ukrainian republic of the Soviet Union (Figure 1 (a) and (b)). It is located about 105 kilometers north of Kiev, capital and major city of Ukraine, with 2.4 million residents. The town of Chernobyl gave its name to the nearby Chernobyl nuclear power station, 15 kilometers to the northwest, which by 1986 had four of the most modern Soviet RBMK-type reactors in full operation, with two more under construction. Three kilometers away from the reactors was the city of Prypyat, with 45,000 people. The Dnieper River flows through the area on its way to the Kiev Reservoir.

The Soviet Program

The initials RBMK are a Russian acronym which translates roughly as "reactor cooled by water and moderated by graphite." It describes one of the two types of reactors the Soviets have built for electric power production, the other being similar to the United States pressure vessel reactor (the Soviet VVER-type). The RBMK type is the older of the two designs. The Soviets developed it themselves from earlier models that had been first used to generate plutonium for weapons, and to produce heat for district heating.

The Soviets have a strong and growing nuclear power program. At the time of the accident, the USSR generated about 10% of the world's nuclear power from 43 operating reactors, a total of 27 thousand million watts of electricity. Under construction were another 36 reactors representing 37 thousand million watts, while at the planning stage were another 34 reactors or 36 thousand million watts. Figure 2 shows the split by type of reactor as of January 1986. There is a big shift in future plants *away* from the RBMK-type of reactor (labeled as "graphite" on the figure for reasons described below) and *toward* the pressure-vessel type of reactor. The shift was planned *before* the accident had occurred, and was a recognition, we believe, that the RBMK reactors were obsolete, and were not economic compared to modern pressure-vessel and modern pressure-tube concepts.

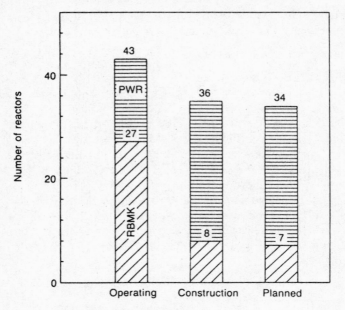

Source: Japan Atomic Industrial Forum

Figure 2 The Soviet reactor program (10% of world's nuclear program)

How A Reactor Works

Before we describe the RBMK reactor and the accident, it is worth-while to review the major parts of a nuclear generating system. One can begin by comparing it with a simple gasoline-powered portable generator that people use in cottages and on trips (Figure 3). The motor burns gasoline, and uses the energy of the hot gases produced to move pistons. Since the pressures from the burning gasoline are high, the engine encloses the "reaction" in strong cylinders. The pistons turn a crankshaft, which spins an electrical generator. The electricity in turn can be used to run lights, work a refrigerator and for other purposes. One controls the power of the engine by a throttle which varies the rate at which gasoline is fed into the engine.

The basic principles are: using a fuel to heat up a fluid (gas), and then using the energy of the fluid to spin a generator. The same principles are used in a nuclear power reactor, though of course the scale is vastly different. The fuel is uranium. Small particles called neutrons split up the uranium atoms; this produces heat, and more neutrons, which keep the reactor going. The heat turns water into steam, and

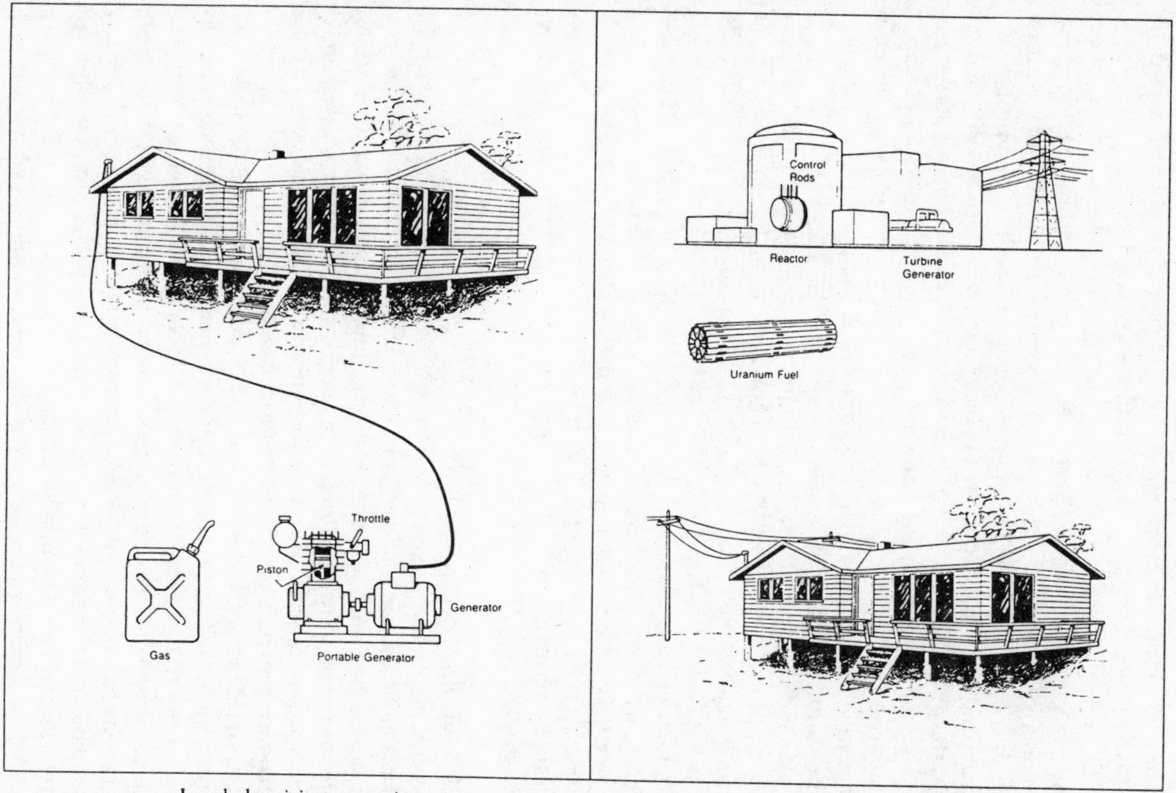

Local electricity generation

Central electricity generation

Figure 3

the energy of the hot steam spins a turbine (which is more efficient than a piston), which in turn spins an electrical generator.

Since the hot steam is at high pressure, it must be kept in a strong container. One controls the power of the reactor by changing the number of neutrons. The power is steady if the number of neutrons produced exactly matches the number used up. If more are used up than produced, the reactor shuts down; if more are produced than used up, the power increases. There are many materials that are very good at absorbing neutrons, for example boron (found in household borax), and these are used in making reactor "throttles." Usually they are formed into rods, and by moving these control rods in and out of the reactor, one can move the power up or down.

We mentioned that a strong container is needed to hold the uranium and the hot water. In the RBMK reactors, the containers consist of about 1,600 small (4-inch diameter) pipes, called pressure tubes. In the VVER type of Soviet reactor, the container is a single huge pressure vessel, containing *all* the uranium and hot water in a large pot. Both concepts are used elsewhere—Canada, Korea, Argentina, Japan, the United Kingdom, India, Pakistan and Italy all have experience with pressure tube reactors of one sort or another, and the United States, France, Germany and many other countries have built pressure-vessel reactors.

How an RBMK Reactor Works
GETTING THE RIGHT CONDITIONS

If one put together blocks of uranium, and tried to split uranium atoms, it would be found that the neutrons moved *too fast* and would miss the uranium atoms too easily. Therefore all of today's commercial nuclear power stations have a way of slowing down the neutrons, by making them pass through a material called a *moderator*. In Canadian CANDU reactors, this moderator is a special type of water, called *heavy water* (it is 10% heavier than ordinary water); in U.S. pressure-vessel reactors, it is an extra supply of the ordinary water which cools the fuel; in the RBMK reactor, it is a solid called graphite. It is the same graphite that is used in a pencil, except that it is purer—both reactor graphite and pencil graphite are forms of carbon, the sort that is burned as briquettes in a charcoal barbecue.

So the heart or *core* of an RBMK reactor consists of a huge container, about as large as a North American house, filled with graphite blocks. The blocks are pierced by about 1,660 vertical holes, in which the pressure tubes and the throttles, or control rods, fit (Figure 4). As

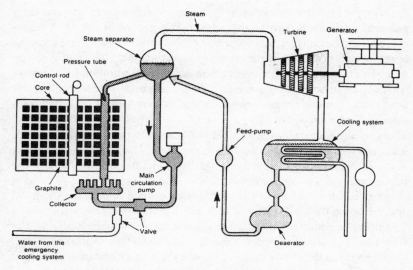

Figure 4 Schematic diagram of the RBMK-1000

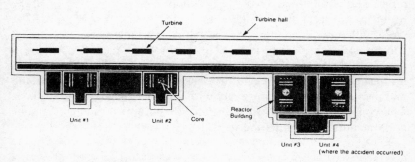

Figure 5 Layout of four reactor units

neutrons split the uranium atoms, the uranium fuel gets hot. Water is pumped from the bottom of the pressure tubes over the fuel. It removes the heat from the fuel, turns to steam in the process, and leaves the reactor core at the top. From there it goes through pipes and gives up its energy to spin *two* large turbines, in an adjacent building (Figure 5). The turbines, in turn, spin electrical generators, and the cooled water goes back into the reactor again. All the reactor itself does is the mundane job of boiling water.

As in all reactors, about 5% of the heat produced by the uranium leaks out to the moderator. In U.S. reactors, the moderator heat is taken away in the steam to the boilers. In the CANDU reactors, where the moderator water is separate from the cooling water, the moderator heat is removed by an independent moderator cooling circuit, consisting of special pumps which circulate the moderator water through coolers, and pump it back into the reactor core—like a transmission oil cooler in a car. The coolers keep the moderator temperature at about 70C, or the same as the hot water in a house. Obviously one cannot do this with solid graphite.

In the RBMK design, the graphite operates at a high temperature— about 700C—and if one could see it, it would be glowing faintly red-hot. This heat flows slowly from the graphite back through the pressure tubes, and is finally taken away by the boiling water. However, the problem with graphite at high temperature is that if it is exposed to air, it will burn slowly, just like the charcoal briquettes on a barbecue. Thus it is *very* important in the RBMK design to keep air away from the graphite. To do this, the Soviets put their entire core in a sealed metal container (Figure 6), and circulate a mixture of inert gases, helium and nitrogen, which do not react with graphite, inside the container. The container was built to withstand the failure of a pressure tube without bursting and letting in air.

CONTAINING AN ACCIDENT

The rest of the structure in Figure 6 is just shielding, to reduce the levels of radiation around the reactor while it is operating. Shielding is used in all reactors so that people can work in the buildings in which the reactors are housed, without getting overexposed to radiation. On the sides of the reactor are shields made of water, sand, and concrete; on the bottom is a concrete shield; and on the top, another concrete shield. *All* the pressure tubes and control rods are attached to this top shield, and it played a key role in the accident. The whole reactor/shield structure is inside a building.

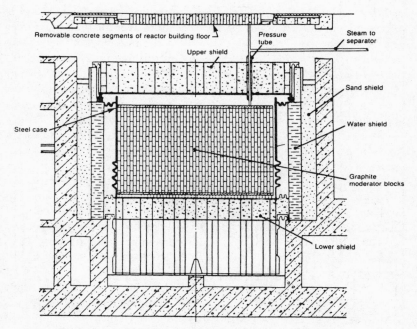

Removable concrete segments of reactor building floor

Upper shield

Pressure tube

Steam to separator

Sand shield

Water shield

Steel case

Graphite moderator blocks

Lower shield

Figure 6 Cross sectional view of reactor vault

In any reactor, if a pipe carrying the water which cools the uranium were to break, and nothing were done, several things could happen:
1) mildly radioactive steam would escape from the pipe and contaminate or damage equipment;
2) since the uranium has lost its cooling water, it could get too hot, and would be damaged;
3) radioactive material normally safely contained inside the uranium could escape to the rest of the plant and to the outside.

This is an unacceptable risk both from a public safety and an economic point of view. To reduce the chances of escape of radioactive material, designers of nuclear reactors normally provide several "lines of defense":

• exceptionally high-quality piping, plus inspection of the piping in-service to see if it is deteriorating. This follows the old adage of prevention being better than the cure.

• normal control systems which, if a pipe break does occur, can shut the reactor down and, in most cases, replace the water that is being lost without damage to the fuel. This *mitigates* an accident after it

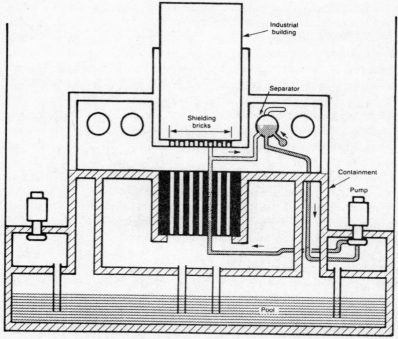

Figure 7 Chernobyl containment

has occurred so that both *safety* and *economics* are respected.

- special *safety systems*, which act only in an accident, and back up the normal control systems. They can shut down the reactor, and replace water as fast as it is lost from any pipe break. (The system which replaces water is called an Emergency Core Cooling system or ECC.) The main function of these special safety systems is public and worker safety, so they are mitigating systems.

- strong leak-tight buildings surrounding the pipes so that even if they do break, and even if radioactive material is released, the steam and the radioactivity are *contained* by the buildings. This does not prevent plant damage, but protects the public by *accommodating* an accident. These structures are normally called *containment* and are a *safety system*.

The Chernobyl unit 4 reactor had shutdown and emergency core cooling, as we will discuss later, but had only a *partial* containment. The pipes *below* the reactor core were inside what the Soviets called "leak-tight boxes." These boxes were connected to a huge pool of water under the entire building—the "bubbler" pond as the Soviets named it. If one of the pipes in the boxes broke, the steam would be forced into the pond, where it and any radioactive particles it contained would be trapped in the water, and the leak-tight boxes would hold. But all the steam pipes *above* the core were inside *ordinary industrial* buildings (Figure 7). Thus if one of these pipes broke, particularly if the break were large, a release of radioactive steam would occur. The amount of radioactivity released would depend on how effective the other systems—shutdown and emergency core cooling—were in preventing damage to the fuel.

This is hard to understand—why not build containment around the whole reactor and all its piping?

Earlier versions of the RBMK reactors, for example the 4 units at Leningrad, do not even have a partial containment. The Soviet philosophy at the time these were built relied on accident prevention and mitigation, and neither their RBMK reactors nor their U.S.-style, pressure-vessel reactors had containment. The accident at the U.S. Three Mile Island plant in 1979 caused a thorough review of safety in all countries, including the USSR. While the Soviets may have already started to add containment to reactors near cities, it is likely that as a result of Three Mile Island, they confirmed that containment buildings were justified at other locations. But, the RBMK is a huge reactor—there is a tall fueling machine at the top that replaces the uranium as it is used up, so the building above the reactor is large—about 71 meters high. The Soviets felt that to put all this in a containment would be difficult and costly. To put the bottom pipes in containment is easier, and this was done. So Chernobyl unit 4 represented a compromise.

The Accident on April 26, 1986

Much of the information on the accident sequence comes from Soviet official sources.[2] The Soviets have published reactor *design* information, in open literature, providing the key characteristics of the RBMK, and sometimes discussing weaknesses with exemplary frankness. Can it be checked? Yes. In Canada, nuclear reactor design is the responsibility of Atomic Energy of Canada Limited (AECL), a Crown Corporation which designed the CANDU reactor, and for which I work as Manager of Safety Engineering. AECL scientists *can* and *have* checked the consistency of the Soviet information on the accident sequence and the design using our own mathematical reactor models, and confirmed that the Soviet information is consistent. We also discovered important omissions from the Soviet presentations in Vienna—they had left some clues, but because there were design weaknesses, they were not highlighted. We have now filled in the blanks, and AECL's interpretation of the accident cause[3] is now accepted by many countries and has been confirmed as plausible by independent assessments done in the United States (by the Department of Energy) and in the United Kingdom, and by the Soviet Union.[4]

How and Why It Happened
A TEST FOR SAFETY SETS IT OFF

It is one of history's ironies that the worst nuclear accident in the world began as a test to improve safety. The events of April 26 started as an experiment to see how long a spinning turbine could provide electric power to certain systems in the plant. What was the reason for the test? Well, the Soviets, in common with many other countries of the world, design their reactors not only to withstand an accident, but also to cope simultaneously with a loss of electric power. This may seem a little strange—to run out of power at an electric generating station—but in an accident the reactor is shut down immediately, so it cannot generate its own power directly. It would normally obtain power from the electric supply *to* the station or from the other reactors at the same site. To ensure an extra layer of defense, it is considered that there is a possibility that these sources have also failed.

The normal backup is to provide diesel engines at the site to drive emergency generators, just as hospitals do in the case of a power failure. These diesels usually start up in 30 seconds, which for most plants is a short enough interruption to keep important systems going. For the Chernobyl reactor, the Soviets felt that this was *not* short enough, and that they had to have an almost uninterrupted supply. Even with the reactor shut down, the spinning turbine is so heavy that it takes a while to slow down, and the Soviets decided to tap the energy of the spinning turbine to generate electricity for the few seconds before their diesels started. The experiment was to see how long this electricity would power the main pumps which keep the cooling water flowing over the fuel.

The test had been done before, on unit 3, with no particular ill-effects on the reactor. However, the electric voltage had fallen off too quickly, so that the test was to be redone on unit 4 with improved electrical equipment. The idea was to reduce reactor power to less than half its normal output, so all the steam could be put into one turbine; this remaining turbine was then to be disconnected, and its spinning energy used to run the main pumps for a short while. At the meeting in Vienna, the Soviets were at some pains to point out that the atmosphere was not conducive to the operators' performing a cautious test:

- The test was scheduled to be carried out just before a planned reactor shutdown for routine maintenance. If the test could not be done successfully *this* time, then the people would have to wait for another year for the next shutdown. Thus they felt under pressure to complete the test this time.

- Chernobyl-4 was a model plant—of all the RBMK-1000 type plants, it ran the best. Its operators felt they were an elite crew and they had become overconfident.

- The test was perceived as an electrical test only, and had been conducted uneventfully before. Thus the operators did not think carefully enough about the effects on the reactor. There is a strong possibility that in fact the test was being supervised by representatives of the turbine manufacturer instead of the normal operators.

How The Trap Was Set

The accident really began 24 hours earlier, since the mistakes made then slowly set the scene that culminated in the explosion on April 26. Table 1 summarizes what the operators did and how the plant responded; here we describe the key events.

At 1am on April 25, the reactor was at full power, operating normally with steam power going to both turbines. Permission was given to start reducing power for the test, and this was done slowly, with the reactor reaching 50% power 12 hours later at 1:05 in the afternoon. At this point only one of the two turbines was needed to take the steam from the reactor, and the second turbine was switched off.

Normally the test would then have proceeded with the next step being to reduce power still further to about 30%. However, the Soviet electricity authorities refused to allow this, as apparently the electricity was needed, so the reactor stayed at 50% power for another 9 hours. At 11:10pm on April 25, the Chernobyl staff got permission to continue with power reduction. Unfortunately, the operator made a mistake, and instead of holding power at about 30%, he forgot to reset a controller and the power fell to about 1%—the reactor was almost shut off. This was too low for the test.

In all reactors, a sudden power reduction causes a quick buildup of a material called xenon in the uranium fuel. Xenon is a radioactive gas,

Table 1   Event Sequence

Time	Event	Comments
April 25		
01:00	Reactor at full power. Power reduction began.	As planned.
13:05	Reactor power 50%. All steam switched to one turbine.	As planned.
14:00	Reactor power stayed at 50% for 9 hours because of unexpected electrical demand.	
April 26		
00:28	In continuing the power rundown, the operator made an error which caused the power to drop to 1%, almost shutting off the reactor.	This caused the core to fill with water and allowed xenon (a neutron absorber) to build up, making it impossible to reach the planned test power.
01:00-01:20	The operator managed to raise power to 7%. He attempted to control the reactor manually, causing fluctuations in flow and temperature. Almost all control rods were withdrawn.	The RBMK design is unstable with the core filled with water—i.e., small changes in flow or temperature can cause large power changes. With most of the rods out, the capability of the emergency shutdown is badly weakened.
01:20	The operator blocked automatic reactor shutdown first on low water level, then on the loss of both turbines.	He was afraid that a shutdown would abort the test. Repeat tests were planned if necessary, and he wanted to keep the reactor running to do these also.
01:23	The operator tripped the remaining turbine to start the test.	
01:23:40	Power began to rise slowly at first.	The reduction in flow as the voltage dropped caused a gradual increase in boiling leading to a power rise.
	The operator pushed the manual shutdown button.	This probably caused the fast power increase due to the rod design.
01:23:44	The reactor power reached about 100 times full power, fuel disintegrated, and excess steam pressure broke the pressure tubes.	The pressure in the reactor core blew the top shield off and broke all the pressure tubes.

but more important it sucks up neutrons like a sponge, and tends to hasten the reactor down the slope to a complete shutdown. As well, the core was at such a low power that the water in the pressure tubes was not boiling, as it normally does, but was liquid instead. Liquid water has the same absorbing effect as xenon. To try to offset these two effects, the operator pulled out almost *all* the control rods, and managed to struggle back up to about 7% power—still well below the level he was supposed to test at, but as high as he could go because of the xenon and water.

It was rather like driving a car with the accelerator floored and the brakes on—it was abnormal and unstable.

Indeed, it is a very serious error in *this* reactor design to try to run with all the control rods out. The main reason is that some of these same rods are used for emergency shutdown, and if they are all pulled out well above the core, it takes too long for them to fall back into the high-power part of the reactor in an emergency, and the shutdown is very slow. The Soviets said that their procedures were very emphatic on that point, and that "Not even the Premier of the Soviet Union is authorized to run with less than 30 rods!"

Nevertheless, at the time of the accident, there were the equivalent of only *6 to 8* rods in the core.

At any rate, the operator had struggled up to 7% power by 1am on April 26, by violating the procedure on the control rods. He had other problems as well—all stemming from the fact that the plant was never intended to operate at such a low power. He had to take over manual flow of water returning from the turbine, as the automatic controllers were not operating well at the low power. This is a complex task to carry out manually, and he never did succeed in getting the flow correct. The reactor was so unstable that it was close to being shut down by the emergency rods. But since a shutdown would abort the test, the operator *disabled* a number of the emergency shutdown signals.

After about 30 minutes of trying to stabilize the reactor, by 1:22am, the operators felt that things were as steady as they were going to be, and decided to start the test. But first they disabled one more signal for automatic shutdown. Normally the reactor would shut down automatically if the remaining turbine were disconnected, as would occur in the test, but because the staff wanted the chance to *repeat* the test, they disabled the shutdown signal also. The remaining automatic shutdown signals would go off on abnormal power levels, but would not react immediately to the test.

Let us digress briefly to examine the state of the reactor. Most of the shutdown signals had been disabled. The control/safety rods had mostly been removed, and the power was abnormally low. As well, the core was filled with water *almost* at the boiling point. We mentioned that liquid water is a good absorber of neutrons. So if it boils suddenly (water being replaced with steam), fewer neutrons get absorbed and the power goes up. In normal operation, this is not a problem as the reactor is designed to cope with this change. But at low power, with the core filled *completely* with water, sudden boiling would cause a rise in power *at a time when the shutdown systems were abnormally slow*. The Canadian analysis by Chan, Dastur, Grant, and Hopwood of AECL and Chexal of the U.S. Electric Power Research Institute (EPRI),[5] showed that this effect by itself would be too small to *start* a bad accident, but it would *accelerate* a rise in power that had already started.

The Canadian/EPRI analysis in fact points to a more fundamental weakness in the shutdown system design. The control rods (which are also used for shutdown) travel in vertical tubes, and are cooled by flowing water. Normally the control rod moves in and out of the reactor to control the power—moving in (adding more neutron absorber) to reduce power and out to increase it. So as the control rod moved in, it would replace the water, and as it moved out, it would be replaced *by* water. The trouble with this scheme is that water *also* absorbs neutrons, so the effect of moving the rod would be small. To enhance its effect, at the bottom end of most of the rods, there is attached *another* rod, made of graphite—called a displacer. Graphite, as stated, does *not* absorb neutrons very well. Thus when a control rod moves in, it replaces not water, but graphite—so its effect on the number of neutrons is larger; and similarly when it moves out (Fig. 8a).

So far so good. The weakness lay in the way this scheme worked if the reactor was *not* operating normally. Just before the accident, most of the control rods were pulled out of the reactor, so far out in fact that even the graphite section was above the bottom part of the reactor—the control rod tubes at the bottom contained only water (Fig. 8b). Even this would not normally matter, because very little power is usually generated at the bottom. But Canadian simulations and the pattern of damage to the reactor suggest that just before the accident, most of the reactor power *was* being generated near the bottom. If the control/shutoff rods were then driven slowly in, the *first* effect would be to replace water (which absorbs neutrons) by graphite (which does not) (Fig. 8c).

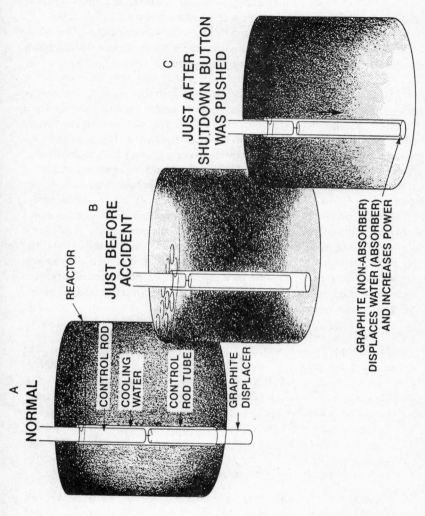

Figure 8 Effect of shutdown rods

In other words, driving in the control/shutoff rods, which was supposed to shut down the reactor, would have precisely the *opposite*
effect—it would cause a fast power *increase* instead.

With this in mind, let us return to the sequence of events. The staff
were now in the worst possible situation for a rise in power which could
not be stopped. And this is what happened.

THE TEST BEGINS

At 1:23:04, the turbine was disconnected and its energy fed to 4 of
the 8 main pumps. As it slowed down, so did the pumps, and the water
in the core, now moving more slowly over the hot fuel, began to boil.
Twenty seconds later the power started rising gradually because of the
effect of coolant boiling, and at 1:23:40 an operator pushed the button to drive in the emergency rods and shut down the reactor. We do
not know for sure why he did this—the individual was one of the early
casualties—but likely he saw either the power begin to rise or the control rods start to move slowly in to overcome the power rise. The shutdown rods began to move in slowly. Our analysis[6] shows that this
attempt to shut down the reactor in fact caused a *large, fast power rise*.
It is acknowledged as plausible by a Soviet paper, presented at a public
conference in October 1987. The paper uses very technical language,
but it is worth quoting.[7]

> As we know, before the accident the excess reactivity in Unit 4 of
> the Chernobyl plant was substantially below that required by the
> rules, and the possibility was not excluded that positive reactivity
> would be introduced in the first few seconds after the AZ-5
> (emergency protection) button was pressed. Analyses with one-
> dimensional models have shown that if the calculated initial vertical
> field varied within limits set by the differences between the indica-
> tions of the various sensors and if the possible errors in these readings
> were applied (up to 25% for reactor power) the positive reactivity
> introduced could vary within the range from zero to 1.5 beta.

Paraphrasing: because the control/shutdown rods were all
withdrawn, pushing the emergency shutdown button could *increase*
power. Because the reactor power was difficult to measure accurately
at low levels, there was some uncertainty as to how fast the power went
up, as measured by a parameter called "beta," which could be anywhere
from zero (no effect) to 1.5 (very fast rise indeed). The Canadian/EPRI
analysis had previously predicted beta to be 0.75—right in the middle
of the range quoted.

In any event, within 4 seconds, the power had risen to perhaps 100 times full power and had destroyed the reactor.

THE TEST ENDS DISASTROUSLY

The power surge put a sudden burst of heat into the uranium fuel, and it broke up into little pieces. The heat from these pieces caused a rapid boiling of the cooling water, and a number of pressure tubes burst under the strain. The steam escaped from the pressure tubes, burst the metal container around the graphite, and lifted the concrete shield on top of the reactor. This broke all the remaining pressure tubes.

Damage to the Plant

THE SAME DAY

The power surge destroyed the top half of the reactor core, the building immediately above the reactor, and some of the walls on either side (Figure 9b). The Soviets commented somewhat ironically that the leak-tight compartments below the reactor survived intact.

Some burning fragments of fuel and graphite were thrown out in the explosion, and landed on the roof of the adjacent turbine building, causing about 30 fires on the asphalt roof and elsewhere. The Soviets' first priority was to put these out, so the damage would not spread to the three reactors operating nearby. Local firefighters had extinguished all the fires by 5am, but at a terrible personal cost: many of them were overexposed to radiation and were among the early casualties.

The destruction was not, of course, a result of a nuclear explosion, but rather of steam and perhaps chemical explosions, so the damage was confined to unit 4. Indeed, unit 3 kept generating electricity for several hours, and the other units for somewhat longer, until they were all shut down in a controlled manner because of the increasing radioactive contamination of the area.

THE NEXT TEN DAYS

The next step was to try to cool off the damaged core. The water pipes had been broken in the explosion, so an attempt to flood the core with water was unsuccessful, and only caused the spread of radioactive contamination to the nearby unit 3. The graphite, meanwhile, had been exposed to air by the destruction, and was being heated by the small amount of heat coming from the fuel which, although broken up, was still in the reactor and piping compartments. By the second day, the graphite had begun to burn in places, as clearly seen in a film taken

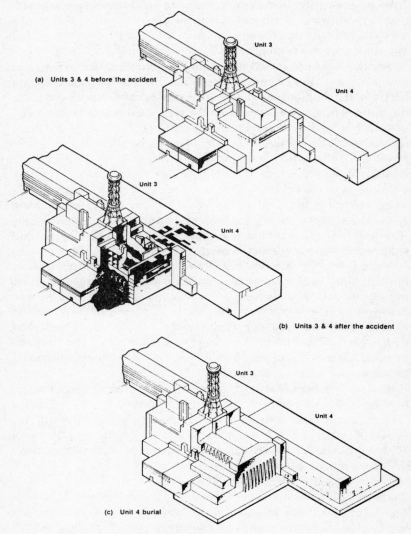

(a) Units 3 & 4 before the accident

(b) Units 3 & 4 after the accident

(c) Unit 4 burial

Figure 9 Damage from the accident

from a Soviet helicopter. Eventually about 10% of it was consumed. The burning was not altogether bad—it caused an air draft through the damaged core that took some of the heat away from the fuel, but the same air was reacting chemically with the fuel and causing it to release radioactive particles. So the Soviets decided to smother the core, and from April 28 to May 2, flew hundreds of helicopter sorties over the reactor, dropping 5,000 tons of mainly lead, sand, clay, and limestone—the idea was that these materials would trap radioactive particles before they could escape.

The materials did shield the core, but like putting a tea cozy over an electric kettle, also trapped the heat, so the fuel began to heat up again. The Soviets solved *that* by pumping nitrogen into the bottom of the core, which really did the job—cooling off the core and putting out the graphite fire.

In fact, after the initial power surge, the fuel never did melt—there was no "meltdown." We know this both from Soviet calculations and from instruments positioned in the reactor itself by helicopters. The reason was the good flow of air, and the ability of the enormous amount of graphite to soak up heat—the same reason why a heavy steel car engine does not melt even though gasoline is burning inside it. The Soviets *were* worried about the possibility of the core collapsing into the water pool below, causing a burst of steam. So they sent courageous divers *into* the pool to open some valves and empty it. In the end, between 2% and 8% of the significant radioactive species of material escaped from the plant, much of them being deposited as dust or particles close by, and the rest being carried by wind over Ukraine, Belorussia and Europe.

The Long Term

With the situation stabilized, the Soviets' next tasks were to remove radioactivity from the site so the other three units could be restarted, and to shield the damaged reactor more permanently. Graphite pieces that had been blown out onto the roof were thrown back into the reactor pit by teams of people working at top speed to reduce their exposure to radiation. Remotely-controlled bulldozers were used to scrape off the contaminated topsoil, buildings were washed down with special chemicals, concrete was poured on the ground to keep down radioactive dust, and deep concrete walls were built in the ground around the site to prevent contaminated groundwater from spreading.

The damaged reactor itself has been surrounded by a concrete "sarcophagus," as the Soviets call it, which shields the radiation sufficiently that working near it is possible. The fuel will continue to generate a small amount of heat for a long time, so fans blow cooling air through the core, and filters remove radioactive particles from the air on its way out. Figures 9 (a,b,c) show the state of the buildings before and just after the accident, and the burial of the damaged reactor six months later.

Why Things Went Wrong—Ideas of Safety

What went wrong? To be sure, the operators made some mistakes. But a mistake should not lead to such disastrous consequences. The problem was that the design was not *forgiving* of mistakes.

The most important weakness was in shutting the reactor down. At Chernobyl, the shutdown rods are part of the normal plant control system, rather than being a separate emergency system. (By contrast, in the Canadian CANDU reactor, the control system can shut down the reactor for most accidents but is backed up by *two, independent, separate, powerful* shutdown systems dedicated only to safety. For a description of the CANDU design and its safety philosophy, see Notes 8 and 9; for more details on the comparison of CANDU and Chernobyl designs, see Note 10.) This meant in Chernobyl that a fault in the control system could *also* disable the emergency shutdown. In other words, *the shutdown effectiveness depends on the reactor being operated properly.* There *has to* be at least 30 (other) control rods in the core for the emergency shutdown to work properly, and the reactor should not normally be run at low power. If these requirements are violated by the operators, as they were in the accident, then emergency shutdown can be slowed down considerably. In fact, because of their design, pulling the rods all the way out not only slowed them down, but caused the power to *increase* when they first started going in, as we discussed.

The second weakness, as we have mentioned, is that Chernobyl had a partial containment. The accident bypassed this partial containment, with the releases all going out through the burst-open top of the reactor core into the region of the building where there were *no* leak-tight boxes. Western reactors have a complete containment which surrounds *all* major cooling piping with a concrete building designed to take the pressure of *all* the steam released in an accident (see, for example, the CANDU 600 design [Figure 10]). Furthermore, in an enclosed containment, the steam and water from a broken pipe will make the

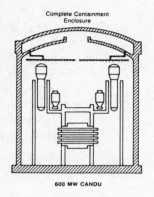

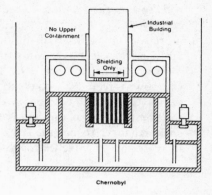

Figure 10 Containment

atmosphere in the building like a rain forest during a tropical storm, and the water will dissolve almost all the radioactive iodine and cesium released. The iodine and cesium, so trapped, will not escape even if the containment leaks. This chemical reaction did not occur at Chernobyl because the hot fuel and graphite were exposed to air, and there was no containment on top of the reactor.

Lessons Learned

The world will continue to study Chernobyl for years to come. Each country with a nuclear power program has been scrutinizing the accident to see what lessons apply to *their* design and operation. In the meantime, the Soviets have drawn certain conclusions for their own program:[11]

a) They recognize how important their operators are. On the positive side, they are installing improved displays of information in the control room; improving operator training; improving procedures; and making it much more difficult to disable safety systems. On the negative side, the importance of procedures will be reinforced— violating one will be a *criminal* offence, and the designers and operators of Chernobyl unit 4 have already been tried and sentenced.

b) On the hardware side, in the short term the Soviets first mechanically prevented control rods from being withdrawn too far in the RBMK reactors. This caused a misshapen power distribution so the reactors had to be run at reduced power. They are now redesigning the rods so that the graphite cannot be lifted above the bottom of the core even if control rods are fully withdrawn. They also required more rods to be inserted—in fact, the plants will now shut down automatically if the operator tries to pull out too many rods. The motors of the shutoff rods have been changed to speed up shutdown—from 20 seconds to 10 (compared to only 2 seconds in CANDU). A new fast shutdown design is being developed. Finally, the composition of the fuel is being changed (more enrichment). These changes reduce the *size* of the positive void coefficient, and increase the effectiveness of the emergency shutdown.

1 The Victims of Chernobyl

Rarely has a topic elicited such controversy as that of the victims—current and future—of the Chernobyl accident. The initial reason for the debate appears to have been the erroneous Western press reports about the first casualties. The figures cited, which have included both 2,000 and 15,000 early victims, are still condemned by Soviet sources, which speak of "scaremongering" tactics, particularly by press agencies of the United States. The consequence has been a natural wariness on the part of the Western press to speculate on the fallout from Chernobyl. Moreover, after the initial week-long silence on the part of the Soviet media, the plethora of articles and the eventual Soviet report to the International Atomic Energy Agency in Vienna received widespread praise in the West.

This reaction appears to have been simplistic. There has been much to commend in Mikhail Gorbachev's policy of glasnost, but it cannot conceal either wide divisions on the topic of Chernobyl within Soviet society, or the desire of the Soviet leadership to communicate the established official view, no matter how many articles are published on various aspects of the topic. Over the past year, a schism has developed in the Soviet Union—and perhaps in the Ukrainian SSR in particular—between the opponents of nuclear power, who seem to come from a broad spectrum of the population, and prominent Soviet scientists who have been assigned the task of ensuring the future of nuclear energy in the Soviet Union.

The latter have developed a wide series of articles and discussions with the ostensible purpose of demonstrating that the impact of the Chernobyl accident was significantly less than initially feared. This may seem a strong statement to make, but it is easily proven. Can one, for example, find an article by a prominent Soviet scientist, written after the Chernobyl event, which attempts to show that the fallout and future casualties are actually much worse (or even slightly worse) than first thought? The answer is no. Yet were the debate to be truly open on the subject, it is clear that such a viewpoint must surely have been expressed, if only out of scientific inquiry. Similarly, as will be seen below, those Western scientists who have forecasted relatively high future cancer fallout from the accident have been condemned roundly by the Soviet side: first, for making predictions in the first place when they have insufficient evidence; and second, and perhaps more important, for causing fear among the local population.

The exception to the above rule has been in commentary about the lives of the participants in the accident, in plays, documentaries and in the works of writers such as Yurii Shcherbak. Glasnost in this sphere does not threaten the prognostications of scientists. It does, however, undermine their general views about the impact of the accident.

Could one say, however, that Soviet scientists' views on Chernobyl represent a sober, realistic type of thinking, as opposed to the more emotional and passionate outpouring of writers and journalists who are more concerned with the suffering of individual families, the rational as opposed to the irrational approach?

Unfortunately, there is evidence to suggest that the figures released on the radiation fallout from the accident, when Vice-President of the Academy of Sciences of the USSR, Evgenii Velikhov, was at the accident scene, were far from accurate. For example, newspaper accounts, such as those in Izvestiya, were obliged to repeat the figures released by TASS, which are now acknowledged by the newspaper's staff to have been inaccurate and misleading. By citing an average figure of 15-20 millirems per hour as the radiation level for the 30-kilometer zone, the Soviet press did not alert readers to the fact that around the reactor itself, the level was up to 1,000 millirems per hour.

Does this mean that the total number of short-term casualties is likely to be greater than the final Soviet tally of 31? Such a statement is impossible to corroborate, but nonetheless is almost certainly the case, even from the most conservative perusal of events. A Samarkand journalist informed the author in November 1987 that the first Uzbek journalist to visit the accident scene remains seriously ill as a result of radiation contamination.[1] Radioactive dust forced the coverage of existing wells holding the water supplies of villages up to 250 kilometers north of the nuclear power plant, according to the Soviet press.[2] Further, the most significant threat to the health of the people in the zone was the lack of attention to the most basic safety devices immediately after the accident.

The Evacuation Controversy

The evacuation of the 30-kilometer zone has been described in detail elsewhere. Here, our aim is to elucidate some of the questions that are still being raised about the first hours and days after the accident. The official Ukrainian government line was expressed by I.S. Plyushch, Chairman of the Kiev Oblast Executive Committee, in April 1987. He maintained that the Government Commission (established by the Soviet government immediately after the event), the Central Committee of the Communist Party of Ukraine and the Ukrainian government worked "around the clock" to ensure that the evacuation occurred as rapidly as possible. This included the 50,000 population of Prypyat and 68 other population points located in the danger zone. People, he stated, were resettled in other regions of Kiev Oblast and provided with temporary accommodation, food and clothing.[3]

Several authorities have suggested that the entire evacuation process was simply a precaution and that the population of Prypyat and other areas was never in any danger from the radioactive fallout. Thus Aleksandr Kondrusev, the head of the main sanitary and epidemiological department of the Ministry of Health Protection of the USSR, stated that the decision was made to evacuate families from many areas well beyond the 30-kilometer zone, "for safety's sake." The radiation levels, he remarked, "gave no cause for anxiety," but the medical authorities simply wanted to ensure the maximum safety.[4]

In the summer of 1987, Leonid Ilyin, the Vice-President of the Academy of Medical Sciences of the USSR, and a key authority on Chernobyl, was asked to comment on the view that the evacuation of Prypyat was somewhat delayed. He responded as follows:

Based on accepted criteria, I can state with full responsibility that the population of Prypyat was exposed to radiation far below the permissible thresholds. We removed these people only because the accident was unusual, complex and because we could not risk people's lives. So there can be no talk of delays.[5]

From a somewhat different perspective, Dr. Robert Gale of the UCLA Medical Center, who performed bone marrow transplants on the first accident victims, also defended the delayed evacuation of Prypyat. In a speech to a medical convention in Washington, D.C. in November 1986, Gale stated that the authorities had delayed the evacuation of the city in the belief that residents who remained in their homes would be shielded from the radioactive fallout. Had the buses to remove people been available before April 27, 1986, people would not have been allowed to board them because they would risk exposure to radioactive particles. Thus the Soviet authorities acted "judiciously."[6]

However, the evidence shows that the inhabitants of Prypyat were exposed to radioactive fallout in the first hours after the accident, simply because they were unaware of what had happened at the fourth reactor unit of the nearby power plant. This is illustrated in several Soviet accounts, of which the following few examples are typical.

In November 1986, the Soviet author Andrei Illesh was interviewed in the weekly *Moscow News*, a newspaper that has been at the forefront of the campaign for glasnost in Soviet society. The reporter noted that in Illesh's book on Chernobyl, no answer had been provided to the key question: what had been carried out in Prypyat, or what the city's authorities had been doing in the first hours after the accident. Illesh responded that he had "failed to discover any sort of consistent line in the city Soviet's behavior." Lamentably, he pointed out, children were playing football in the streets on the morning after the accident and a wedding party had been held in the city on the same evening. Yet the Prypyat city government did have some notion of what had occurred, but were "stricken by anemia."[7]

During the trial of the nuclear plant's Director and Chief Engineer in Chernobyl in July 1987 (see Chapter 3), one of the key charges against V. Bryukhanov, the Director, was his failure to evacuate his staff immediately. Workers had arrived for the next shift at 8am on April 26, 1986—the main reactor fire having been extinguished only by 5am—and had lingered around precariously. In Prypyat, it was noted in a report on the trial, the Saturday in the city "was the same as any other." Children were playing outside, people went fishing, all the stores were open. According to Andrei Pralnikov of *Moscow News*, the general response from "knowledgable people" to the question why

Prypyat citizens were only evacuated 36 hours after the accident was either that radiation levels were only rising gradually, or that buses and trucks were assembled in the shortest possible time. These responses, stated Pralnikov, revealed an "absence of glasnost."

The correct comment on the question, he continued, was provided by Yurii Grigoryev of the Institute of Biophysics of the Ministry of Health Protection of the USSR, who stated that the evacuation should have been carried out earlier and that people should at least have been informed of the danger over the radio.[8] However, as Lyubov Kovalevska, the former editor of the Prypyat newspaper *Trybuna energetika*, noted, no word was broadcast on the local radio until 12 noon on April 27, which was almost 35 hours after the accident occurred. Then, without explanation, the citizens were informed only that they were to be evacuated for three days![9]

At the Chernobyl trial, of which we have only the briefest of accounts, as a result of the prohibition of the public from all but the first and last days, it transpired that once the public had been informed about the accident, they were quickly misled about the radiation levels. People were actually informed that the radiation situation in Prypyat and areas around the plant was "favorable" in the first days after the explosion. "They remained not far from the destroyed block, and got a considerable dose of radiation." What was the consequence of such misinformation? Only that when more accurate figures and accounts were issued subsequently, no one believed them.[10]

The scene that emerges is of a few individuals trying to draw attention to a perilous situation, while the majority of authorities, from the plant's Director to the local party and government authorities acted as though nothing unusual were happening. Yurii Shcherbak received letters from Prypyat workers who had called the authorities to ask why instructions had not been given to keep children indoors. They were informed that it was none of their business and that "the decision will be made in Moscow."[11] According to a report in *Yunost*, senior officials in Prypyat on April 26, ordered the schools and stores to remain open.[12]

In September 1987, the Ukrainian journalist, Vladimir Yavorivsky, wrote a bitter account of the events in Prypyat in the immediate aftermath of the disaster in the weekly *New Times*, which in the past has hardly been perceived as one of the vehicles for glasnost. According to Yavorivsky, a party meeting was held in the city and furious accusations were made against the local party leaders who had refused to countenance an evacuation. One rank-and-file member evidently shouted that he would spend the rest of his life trying to determine who was responsible for the decision to leave children and "our families" in the "contaminated city." It also described how thousands of youngsters were forced to walk the streets of Kiev during the May-Day parade simply to give an impression of normality. Yavorivsky noted that the local authorities had been warned of a possible accident by Kovalevska's March 27, 1986 article in the weekly *Literaturna Ukraina*, but that the young Prypyat journalist had almost been fired for writing the article—"she was saved by the accident—however bitterly ironic that may sound."

Yavorivsky's article, which effectively demythologizes the official account of a smooth problem-free evacuation process, was published only 18 months after the disaster, which reflects the slow progress of openness in this particular sphere. Were the local party authorities aware of the extent of the accident? This is a moot point. Yavorivsky comments that Bryukhanov had informed the Kiev Oblast civil defense troops that they were dealing only with a fire on the roof of a reactor, not radioactive fallout. But the former plant Director may have been something of a scapegoat. Certainly Moscow was aware of some details (but perhaps not the entire scope) of the accident. According to the editors of *Izvestiya*, there was a furious debate among the editorial board over whether a story should be published on the day after the accident (the newspaper is issued daily), which could hardly have been the case had there been no news to report.[13]

If Prypyat residents had known of the danger immediately, would they have been able to take preventive measures? According to Yurii Grigoryev, several accident scenarios had been rehearsed before April 1986, including one that anticipated the fallout of a large quantity of radioactive iodine. A textbook on the topic had been prepared by the Moscow doctor Angelina Guskova, but because of "bureaucratic holdups," the book had not been published by the time the accident occurred.[14] Nevertheless, in the article in *New Times*, Yavorivsky describes how "fearless nurses," on their own initiative, went from house to house delivering potassium iodide tablets. An early warning on the radio, therefore, could have done much to alleviate the situation.

There were other disconcerting factors about the evacuation process. The first is that party affairs were given priority. The families of Prypyat City Party Committee members were the first to be evacuated, and the local party committees were quickly relocated. The Prypyat City Committee was rehoused in the town of Poliske, 50 kilometers to the west, while the Chernobyl Raion Party Committee (also a subordinate body of the Kiev Oblast Party Committee under Hryhorii

Revenko) was reestablished at "full strength" in the headquarters of the Borodyanka Raion Party Committee "on the very next day after the evacuation."[15] A number of plant operatives from the damaged Chernobyl unit were transferred directly to construction work on the Khmelnytsky nuclear power plant, at Netishyn, in western Ukraine.[16]

In contrast to this relatively quick transfer of location for party and plant operative workers, children were not removed from the zone until mid-May, and the evacuation process on the Ukrainian side of the danger zone, which preceded that in Belorussia, was basically completed only by May 21, almost four weeks after the accident occurred.[17] Soviet writers have stressed frequently that the decisive event in the history of the evacuation process was the arrival in Chernobyl of CC CPSU Politburo members Egor Ligachev and Nikolai Ryzhkov on May 2. Yet it took a further two weeks before the decision to end the school term was taken. Again, one assumes that the overriding factor was the desire to minimize the impact of the accident.

A further factor that is worthy of discussion is the direction of the evacuation. The path of the radioactive cloud in its umbrella movement north and northwest of the Chernobyl plant has been clearly delineated. Nevertheless, the main body of evacuees, 23,000 citizens of Prypyat, were moved to Poliske, which appears to have been in the direct line of fallout. It is almost inconceivable that this route avoided danger. In fact, in March 1987, when scientists were examining the water supply of the area for contamination after the spring flooding, Poliske was frequently a site of the main investigations. The question arises whether the boundary of the 30-kilometer zone coincided with the boundary between Chernobyl and Poliske Raions. This could have been the case because the number of villages evacuated in Ukraine—65—was almost the same as the officially reported number of villages in Chernobyl Raion.[18] By adhering to official boundaries, the local authorities may have compounded the situation. Eventually, villages in the adjoining Poliske Raion also had to be evacuated.

How many were evacuated? The official total varies from 116,000 to 135,000, but omits the children, who were evacuated from cities as far north as Gomel and from Kiev in the south, which raises the total to over 500,000. Officially 90,251 persons were evacuated from the Ukrainian side of the zone, and 18,000 from the Belorussian side to the north. A Foreign Ministry spokesman, Boris Pyadyshev, reportedly stated in September 1986 that a small number of people had been removed from the Bryansk region of the Russian Republic, which would account for the small discrepancy in the official totals.[19] As noted, however, the evacuation was a delayed, disjointed and unhappy affair that is still being debated in the Soviet Union. Had ample warning of the danger been provided in good time, the number of victims, fatal or nonfatal, would have been reduced significantly.

Leonid Telyatnikov, the fire chief at the plant, remarked how in hospital with severe radiation burns he "was told unimaginable things which happened after the accident in Pripyat. After that I stopped respecting the city leaders."

The First Casualties

Were the Soviet medical authorities prepared for an accident of the nature of Chernobyl? There is evidence that they had been studying the biological effects of radioactivity on the body for at least four decades.

Yurii Grigoryev, cited above, had been at the forefront of a "unique experiment" designed to evaluate the effects of various doses of radiation on the body. The experiment, which appears to have been a serious preparation for the irradiation of Soviet cosmonauts from cosmic rays, was conducted on 250 dogs over several years. The dogs had been irradiated from gamma sources, after which they were observed every two months and the scientists evaluated the status of their bone marrow, their immunological and biochemical reactions, their higher neural effects, and their reaction to the various doses. After the irradiation process was completed, the dogs and their descendants were observed for the rest of their lives. The process lasted sixteen years. According to Grigoryev, the main result was the relative ineffectiveness of the irradiation process: malignant tumors were no more frequent than would normally be the case and the dogs' reproductive capacity was not impeded.[20]

Such experiments were considered necessary because the cosmonauts, as a result of sun flashes and travel in the earth's radiation belts, may receive up to 100 roentgens per year of irradiation. The figure is four times higher than the maximum amount officially allowed for those cleaning up the Chernobyl site after the accident, but Grigoryev pointed out that the permissible doses are calculated with a large surplus in favor of man. Twenty-five roentgens, he noted, is considered safe by all international commissions, with the provision that for those people subjected to such a dose, the total dose for the rest of their lives should not exceed 50 roentgens.

In the USSR, explained Grigoryev, the average background dose of radiation accumulated in a person's lifetime is about 25-28 roentgens. In the past, these levels were exceeded considerably by x-rays of the ribcage and gastrointestinal tract, but today improvements in technology enable significantly lower doses through x-rays. However, the Chernobyl event subjected its victims to significantly higher doses of radiation than those received by cosmonauts. The apparent optimism of Grigoryev and his associates concerned the animals' resistance to relatively low levels of irradiation.

According to Valerii Legasov, the First Deputy Director of the Kurchatov Institute of Atomic Energy of the Academy of Sciences of the USSR, the accident took place in the one area where it was expected the least, namely, a nuclear power plant. He had been entrusted several years earlier to lead a group concerned with industrial safety, particularly the potential dangers from various types of energy installations. Because of the dramatic increases in energy capacity, and the concentration of this capacity in a small area, Legasov was quite prepared for a major accident, but not in the nuclear sector. In any event, he noted in the interview, it was necessary to carry out illegal actions consistently at the fourth reactor unit in order to bring about the accident situation.[21] In short, in theory the accident should not have occurred, or rather the authorities were not prepared thoroughly for the Chernobyl accident.

Soviet reflections about the measures taken after the accident indicate both a sad pride and an ambivalence about the methods adopted to deal with the first casualties. In a rare article in the press, Angelina Guskova, the Director of Moscow's Hospital No. 6, where the most seriously injured victims were treated, noted that she is now "considerably older and wiser." In the tense days after the event, she noted, the "young assistants" of the hospital "showed their best side." Some "tens" of people, for whom there appeared to be little hope, were brought back to life, but she experienced a sense of "boundless sorrow" when several of the sick did not survive.[22]

In April 1987, the Novosti press agency published one of the most detailed accounts of the treatment of the first victims and debated the effectiveness of the bone-marrow transplants carried out by Guskova's team, with the assistance of the U.S. medical squad led by Dr. Robert Gale.

Immediately after the accident, it stated, some 299 patients were taken to Soviet clinics, having been diagnosed as suffering from "acute radiation sickness." Of this number, clinical, instrumental and

laboratory tests reportedly confirmed the diagnosis in 237 cases. According to the Novosti report, 139 of this number were said to be "at the
first stage of the disease," and after a stay in hospital, the majority were
able to resume work under close medical observation.[23] According to
a Western source, these victims would have received up to 200 rems
of irradiation, a potentially very harmful level.[24]

The remaining 98 patients were diagnosed as follows: 53 had
received the second degree of radiation sickness (200-400 rems); 23
the third degree (400-600 rems); and 22 the fourth degree (600-1600
rems). Of the 29 who died officially of radiation sickness (in addition
to the two killed during the accident itself), 21 were from the fourth
degree category, 7 from the third degree, and 1 from the second degree:

> Regrettably 28 (?) patients died despite all kinds of treatment, from
> bone-marrow transplant, haemo- and plasma absorption and
> plasmophoresis to cryogenization of their own thrombocytes read-
> ministered to the body...Yet the patients were so badly affected by
> radiation and their skin and internal organs were damaged so irrever-
> sibly that it was hard to expect recovery. Thus 14 patients had burns
> affecting 80-90% of their skin.[25]

The Moscow patients had suffered serious radiation exposure of their
bone marrow, blood, skin, connective tissue, stomach, intestine and
lungs. They required the services of various Soviet specialists, and "marrow transplant experts" assisted the Soviet team in 30 operations. The
success of these operations varies according to the source consulted.

Robert Gale maintains that there were not 237, but 500 patients,
which is an astonishing divergence from the official Soviet figure,
although he may have included also those victims sent to hospitals in
Kiev. During an interview with the Canadian *Maclean's* magazine, Gale
stated that during such accidents, a small fraction of people suffer
irreversible damage to their bone marrow. Of the 13 patients in this
category who received transplants, 11 died. The high death rate was
caused in part, in Gale's view, by the fact that accidents with Soviet
reactors, in contrast to North American reactors, are accompanied by
a major fire: "A marrow transplant can only prevent you from dying
of bone-marrow failure—it cannot prevent you dying of burns or radiation damage to the liver." Moreover, in his view, the overall success
rate was over 90%.[26]

Dr. Gale's views on the success of the transplants are not shared by
Guskova, who has maintained that such transplants "did more harm

than good." In the Soviet view, the marrow transplant cannot be a "panacea for all." Thus if the radiation exposure is in excess of 800 rems, the body's organs and systems would be damaged irreversibly and there would be no point in conducting a transplant. If, on the other hand, the exposure is under 600 rems, then the patients' own bone marrow would breed new cells which would then reject the alien tissue. Consequently, bone marrow transplants, in the Soviet view, can be effective only for patients that have incurred radiation doses of between 600 and 800 rems.[27]

In the case of the 98 seriously injured victims noted above, therefore, they could have been used for between 0 and 22 cases. The actual total was 13. And in fact, the 7 patients in whom a transplant could have been successful died 9 to 19 days after the operation from damage to the skin and intestine caused by radiation, and the remaining 6 patients revealed only temporary or incomplete transplant grafting because their own immune systems rejected the transplant. The Soviet view is that even for those patients who received between 600 and 800 rems of irradiation, a transplant can be dangerous because it may bring about a secondary disease, a fact that has evidently been discovered relatively recently.[28]

Turning from the medical treatment of victims in Moscow to the number of victims generally, the official Soviet figures were disputed by a former Soviet engineer, Igor Gerashchenko, who testified before the U.S. Commission for Verification of the Helsinki Accords in the spring of 1987. Gerashchenko was previously a senior engineer at the Thermophysical Institute in Kiev, but was dismissed in 1980 for his work on human rights. Gerashchenko commented that: "As far as I know from my friends who work in the 2 largest Kiev hospitals, 15,000 individuals died in 5 months in those hospitals alone." He also stated that he was informed by friends at those hospitals that radiation sufferers were recorded as victims of "vascular atonia." Concerning foreign visitors, Gerashchenko maintained that they were shown "Potemkin hospitals," i.e., hospitals that did not contain Chernobyl victims.[29]

The Soviet response to Gerashchenko's claim was predictably swift. On April 1, Radio Moscow's World Service referred to the "outrageous anti-Soviet action" of a certain Igor Gerashchenko, noting that Dr. Gale had recently confirmed the figure of 31 deaths. On the following day, Leonid Ilyin appeared at a press conference and dismissed the 15,000 deaths figure as "utter lies." Ilyin stated that he did not know what motives drove Gerashchenko, but that his figures were "lies and again, lies." The victims, he continued, were concealed neither by the doctors nor by the Soviet government. Not a single case of radiation sickness had occurred among the population of the raions around the nuclear plant, he declared, and the number of accident victims had not grown.[30]

The truth of the matter appears to lie somewhere between the two views. Gerashchenko's figure appears not only impossibly high—the radiation fallout would have been considerably higher in Scandinavia and other countries, for example—but is not based on any firm evidence. The engineer cannot be criticized for failing to cite definite sources from his position as an emigre, but the unfortunate fact remains that since he was not in a position to corroborate his testimony, he should not really have been making the allegations in the first place.

However, as noted above, there is little reason to accept the official figure of 31 deaths as a firm and valid tally. This is made evident from the tragic death of the Ukrainian film director, Volodymyr Shevchenko, who fell ill during the making of a documentary entitled "Chernobyl: A Chronicle of Difficult Weeks" at the plant site between May and August 1986. His death in March 1987 was noted by several Soviet sources, one of which—the weekly *Nedelya*—stated explicitly that Shevchenko died of radiation sickness. Shevchenko's name, however, has never been added to the official list of victims; nor has that of Vladimir Rudakov, a senior official of the Ministry of Medium Machine-Building of the USSR who played a key part in the cleanup operations, and died after a severe illness in January 1988.[31] Moreover, it would be very surprising given the nature of the cleanup operation (see Chapter 5) if these two were the only "unofficial victims" of the Chernobyl disaster. The careful observer, however, is obliged to state only that the number of current victims of Chernobyl stands at more than 33.

Of the early victims, a majority were firemen. Many were on the "first watch" that fought the inferno from approximately 1:30 to 3:30am on the morning of April 26, 1986. In the middle of this period, at around 2.30am, fire units from Prypyat arrived. Then an hour later, units from the fire service of the Kiev Oblast Department of Internal Affairs reached the station, headed by Vasilii Melnyk, which had the task of extinguishing the fire at its final stage. During the first post-accident days, several firemen succumbed to the effects of radiation and the fire in the Moscow hospital, including Viktor Kibenok, Volodymyr Pravyk, Vasilii Ihnatenko, Nikolai Tytenok, Volodymyr Tyshchura and Mykola Vashchuk.[32]

The official death toll from the accident had risen to 31 by the Fall of 1986. By mid-September, Guskova noted that 2 patients remained in hospital and were undergoing plastic surgery to correct skin deformities. A further 15-20 patients were said to be still under observation, while the remaining 129 that had been hospitalized had either returned to work or would do so in the near future.[33] By the spring of 1987, Leonid Ilyin stated that radiation sickness had been diagnosed in 237 people, of whom 209 had been "cured," although 13 patients remained invalids and some of them would undergo operations. Of the 196 who were now working again, none were permitted to return to jobs in the nuclear power industry because they cannot risk further exposure to radioactivity.[34]

Concerning the 31 official victims, the Soviet side has provided over the course of some 20 months, about 20 of the names. It has not, however, released a list of the 237 who were diagnosed as being seriously ill from radiation sickness, who will no doubt be monitored regularly for the rest of their lives. As the vast majority of Chernobyl victims will die in the future, it is the system of future monitoring that is of most interest and concern.

The Monitoring System

In July 1986, a U.S. doctor from the City of New York Medical School, Jack Geiger, who had visited the Soviet Union in the previous month to talk to doctors treating the Chernobyl victims, revealed that the Soviet authorities were to create a new national center to monitor the patients. He noted in the *Journal of the American Medical Association* that all evacuees had been issued with special cards in order that medical data about them could be compiled in a central registry. (Presumably these cards would be in addition to the regular medical cards required by every Soviet citizen.) Tests were to be conducted on the victims' thyroid glands and immunity systems and a comparison made with a group not subjected to high levels of radiation exposure.[35]

The new Center, created in June 1986, was officially announced in the Soviet press in early September 1986. It is an All-Union Scientific Center of Radiation Medicine of the Academy of Medical Sciences of the USSR, composed of three institutes. The Institute of Clinical Radiology studies the influence of low-level radiation upon people's health. The Institute of Experimental Radiology examines scientific data about the effects of various doses of radiation, while the Institute of Epidemiology and Prophylaxis of Irradiation is concerned with the spread of various doses of radiation to various groups of the population and the elaboration of prophylactic measures. Kiev was selected over Obninsk and Gomel as the site for the new Center and, somewhat surprisingly, the controversial Ukrainian Minister of Health Protection, Anatolii Romanenko, was appointed Director.[36]

Radio Kiev announced the official opening of the Center on October 1, 1986, although it is only partially complete. During the 12th Five-Year Plan (1986-90), it is planned to build a specialized clinic, a laboratory-experimental building, and a computing center for the new clinic.[37] It would appear therefore that at the time of its inception, the Center did not possess a computing system that would enable the development of a databank on the progress of the victims of Chernobyl. Just over a week after the official opening, Director Romanenko appeared on Ukrainian television and was interviewed at length in the Ukrainian press about the functions of the new Center.

Romanenko denied that one function of the Center is to "cure people afflicted by a large dose of radiation." He attributed such "rumors" to "voices abroad" that, he declared, had already done more than their share of talking and had even referred to Kiev as a dead city. "With all the responsibility of a doctor," he continued, "I may say that no changes whatsoever occurred that can be reflected in the health of the people of our republic who were not involved directly in liquidating the consequences of the accident." The purpose of this somewhat unnecessary tirade was evidently to emphasize the Center's purpose as a monitoring rather than actual medical institution. It is clear, however, that it would have little function if "no changes" occurred in the organisms of the victims as a result of irradiation.[38]

A main purpose of the Center is to note the effect of small radiation doses on the health of the population. The Institute of Epidemiology has created a register, which reportedly includes the names of those who sustained "even an insignificant dose" of radiation. All information derived from medical inspections of the population is to be placed in a computer bank. Romanenko also elaborated on the current "thresholds" for radiation exposure in the Soviet Union. In his view, the normal borderline for the general public, "which does not affect human health," is 0.5 rems annually, which would amount to 35 rems over the course of an average lifetime. For those people working with radioactive substances, the threshold is 5 rems, which would mean about 150 rems over an average *working* lifetime of 30 years. The figure is significantly less, of course, than the 25 rems noted by Grigoryev above, and well below the maximum norms assigned for the cleanup workers at the Chernobyl site and in the 30-kilometer zone. But would the receipt of low doses bring on cardio-vascular diseases, lung inflammations or stomach ulcers, i.e., are those people who received low radiation doses in any danger as a result?

Romanenko answered this question guardedly and inconsequentially with the statement that in such cases, actual exposure to radiation has less effect than psychological and stress factors that apply to those who either lived close to Chernobyl or worked in the zone of increased irradiation. As will be shown below, the impact of fear of radiation exposure has been so marked among the Ukrainian population that Romanenko's comment tells us very little. He did state, nevertheless, that one of the goals of the Center is to ascertain how the human organism reacts and adapts to higher doses of radiation, "to raise the curtain above these secrets." In his view, the main problem is that whereas one person will become ill with an immediate dose of 100 rems, a similar illness may not appear in a neighbor until he has received 200 rems.

Moreover, one of the tasks of the Institute of Experimental Radiology is to comprehend the mechanism that helps to accelerate the withdrawal of radioactive substances from the body. Romanenko maintains that this occurs especially rapidly in children. Thus, he stated, although children feel the effects of radiation more than adults, the child's organism will eject the harmful substances much more rapidly than that of the adult. Thus whereas the period for the withdrawal of radioactive cesium is 60 days in adults, it is only 9 days for newborns. The statement has been borne out by other sources.

Less satisfactory to the Western observer, however, is Romanenko's blanket statement that all the children from the 30-kilometer zone were observed diligently "and there were no changes to their health," or his comment that "in my opinion there is no threat to health from the increased dosage of radiation." The Minister appears to combine scientific statements with remarks that come close to propaganda. Obviously there is a threat to health from any increased dose of radiation. The Center's purpose is to monitor the health of the victims for their own benefit as well as for the purposes of scientific experiment. Two months later, Romanenko chose his words more carefully during an interview in *Pravda*:

> I am far from thinking that all the danger is now past. It is still of the utmost importance to continue rigorous monitoring of environmental conditions, the purity of foodstuffs and water...The Center still has a considerable amount of work ahead of it.[39]

As for the children, the monitoring process must have been severely hindered by the dispersal of children throughout the country in the first weeks after the accident.

There is also ambiguity over the number of cases studied at the Center, and the number of people examined medically for the effects of radiation in the 30-kilometer zone around the reactor and the areas immediately beyond. Oleg Shchepin, Deputy Minister of Health Protection of the USSR, stated in April 1987 that as a result of the concern of people located at considerable distances from the nuclear power plant, dosimetric and other examinations were carried out, often for purely psychotherapeutic purposes. In other words, many more people were examined than was strictly necessary. However, the total figure provided by Shchepin for all examinations is 80,000,[40] which is less than the official number of evacuees. If one included children, and Romanenko certainly does in his remarks, then perhaps one in seven evacuees were examined—the figure could be considerably less if the outlying areas made up a large proportion of those who were examined because of their anxiety.

In the summer of 1987, however, the Soviet news agency TASS stated that examinations ("complex medical observations") had been carried out on 800,000 residents of Ukraine, Belorussia, Moldavia and the RSFSR.[41] Earlier in the year, the same source, while pointing out that the structure of the new Center envisaged "practically everything" for the systematic observation of the state of health of citizens who resided in the zone of Chernobyl nuclear power plant, stated that about 1,000 people had been examined at the Center by radiometric control and "precise tests."[42] This suggests that either the vast majority of tests were carried out outside the Center, or that between February and April 1987, when Shchepin made his statement, the number of examinations must have risen dramatically from 1,000 to 80,000.

According to Leonid Ilyin, "hundreds" of medical brigades were created. Doctors, nurses and laboratory technicians were assisted by specialists from scientific-research institutes and students of medical institutions. The two main examinations were of the status of the blood and the thyroid gland, and it was necessary to repeat the analysis every 7-10 days.[43] Interviewed in *Pravda* on April 23, 1987, Ilyin noted the establishment of a register for patients (this does not necessarily signify a computerized list, which does not seem to have been available in 1987), and stated that "nearly 100,000" people were listed. The 100,000 had been divided into groups and each group would be observed in detail about once a year. A month later, at a special five-day conference of the IAEA and the World Health Organization in Vienna, A. Moiseev of the Ministry of Health Protection of the USSR stated that the health check would involve 100,000-200,000 radiation victims over a period of 50 years.[44]

Over the course of 1987, it has become increasingly difficult to obtain information about the status of Chernobyl victims. Robert Gale, who has visited the area several times since performing the bone-marrow transplants on the first victims, has stressed that he would like to share Western resources and experience with the Soviet side, even though the primary responsibility for the monitoring process will be that of the Soviet authorities. However, various offers of assistance from the American side have been spurned:

> ....to date, the Soviets have declined all offers from various American bodies to collaborate with them. Our choice is either to find alternatives or to say that we're not working with them.[45]

In addition to this apparent Soviet unwillingness to permit Western assistance—which is perhaps not unusual in that the Kiev Center is carrying out many other functions than monitoring of the victims—the Director of the Center, Romanenko, has been less than forthcoming about the progress of those monitored. Writing in the Ukrainian press in June 1987, for example, he noted that tens of thousands of people had been examined for the radionuclide content of their bodies:

> Analysis of the measurements shows that the irradiation level of the population in the accident zone is significantly lower than the "accident" norms accepted by the majority of national and international organs of radiation safety. For the overwhelming majority of the raions adjacent to the zone, and for the population of the republic, they are comparable to the doses of the natural radiation background.[46]

This sort of blanket statement tells us very little and is reminiscent of the first Soviet figures on radiation fallout, whereby the figures for various parts of the 30-kilometer zone were simply averaged out, giving the reader little idea of where the main danger areas lay. If the fallout is measured for the whole of Ukraine, as Romanenko mentions, then inevitably the average figure falls correspondingly.

An example can be provided: in an interview with *Pravda* on March 22, 1987, Romanenko made a similar statement, declaring that during the first year after the accident, 99.9% of the population did not exceed the usual norm of radiation dosage. It was not clear whether he was referring to the zone around the reactor—he mentioned 1,500 pregnant women in the area—or to the entire republic. In either case, however, his figures lead to bizarre conclusions. Thus if he was referring to (to use the maximum figure cited) 200,000 persons examined after the accident, then this would entail only 200 sufferers from radiation, whereas we have been informed above that at least 237 people were diagnosed as having "acute" radiation sickness.

On the other hand, if one applies the 99.9% to the entire Ukrainian SSR, which Romanenko has done regularly hitherto, then one is left with approximately 50,000 people who exceeded the normal radiation dosage. This figure appears more plausible, but given the tone of Romanenko's remarks, it can hardly have been his intention to call attention to such a figure. Romanenko himself has further clouded the issue by noting that those who worked in the area around the fourth reactor in the first days after the accident received higher doses.[47] This would suggest that the cleanup workers suffered more than the evacuees from radiation sickness (see Chapter 5).

But how many cleanup workers were there? Romanenko does not say. Evgenii Velikhov, the Vice-President of the Academy of Sciences of the USSR, who visited Washington, D.C. in January 1987, provided a figure of 50,000, although this has never been verified by other sources. If accepted, nonetheless, one already has the maximum figure of increased radiation fallout for the entire Ukrainian SSR, as noted above. Moreover, the cleanup crew is not stable, but is constantly changing as workers attain their "rems." For example, the new Director of the Chernobyl plant, Erik Pozdyshev, who was not involved directly in the cleanup work, was forbidden by doctors to remain in the zone (he was replaced by Mikhail Umanets) in the spring of 1987 because he had already attained the maximum amount of radiation permitted.[48] Since such replacements would be occurring all the more frequently at the "lower" end of the scale, the figure of 50,000 is not stable, but is rather the number of people involved in cleanup work at any one time.

Further, are the cleanup workers among those being monitored at the Kiev Center? There is no indication that this is the case. Romanenko, Shchepin and Ilyin, who have been the main spokespersons on the Center, have referred only to the examination of evacuees and the "population of the raions."[49] When cleanup workers are examined at the end of a working day, they are inspected at a local medical center, as might be expected. Once they attain the maximum level of radiation, they are sent home. For all these reasons therefore, Romanenko's comments appear suspect. The Center of Radiation Medicine, as a data bank of the victims, and as the location of a major experiment on the effects of low radiation doses on the population, will take on a status similar to that of Hiroshima and Nagasaki as far as scientists are concerned. It is to be hoped, however, that Romanenko will eventually provide accurate and detailed information about the status of the victims.

Pregnancies

After the Chernobyl accident, a number of pregnant women had abortions rather than risk continuing their pregnancy.[50] The number of such abortions has never been provided by the Soviet authorities, but there was reportedly widespread fear, even well outside the danger zone around the reactor, that the radiation fallout would lead to the birth of abnormal children. One recent emigre from the Soviet Union provided one of the most graphic illustrations of what must have occurred in many areas of Ukraine and Belorussia after Chernobyl:

My sister works as a nurse in a Kharkov women's dispensary. She told me that since the Chernobyl catastrophe, many young women have been having abortions for fear of giving birth to abnormal children. Even women from areas outside the danger zones were afraid they might have been contaminated by radiation in their food, or in their drinking water. Although doctors in the dispensary appealed to the maternal instincts of the pregnant women and tried to persuade them that it was their patriotic duty to have the baby, many of those summoned to the clinic for a follow-up visit turned out to be no longer pregnant. Their excuse would be that they had begun to bleed, and then bleeding turned into a regular menstrual period, and since they felt fine, they didn't bother to go back to the doctor. Naturally, the doctor would realize what had happened, but would be unable to prove it. If a woman gets an abortion on Friday and returns to work on the Monday, no one is any the wiser.[51]

Were such abortions necessary? Dr. Gale maintains that several hundred Chernobyl babies were at risk of abnormal brain development and mental retardation. He noted that in the case of Hiroshima and

Nagasaki there was a direct correlation between mental retardation and a relatively low dose of additional radiation. Another doctor from the University of California at Los Angeles, Dr. Stephen Zamenhof, stated during an interview that "Brain damage could occur in almost every fetus exposed to radiation." Because of the damage caused to the DNA, nerve cells in the brain are unable to divide, the number of neurons is less, and the brain is smaller than normal, a process that occurs during the first 18 weeks of pregnancy.[52] Two Western doctors therefore, while they may differ on the actual figures involved, both concur that there will be a significant number of birth defects as a consequence of the disaster.

According to the study of the Chernobyl accident by the U.S. Nuclear Regulatory Commission, the "maximum collective external" dose of radiation received by the 135,000 evacuees was about 12 rems per person. The report estimated that unless therapeutic abortions were performed after the accident (which was not the case), then about 5% of fetuses in the 8-15 week period could be born mentally retarded. It notes that typically there would be about 300 pregnant women in this critical stage of pregnancy in a population of 135,000. Of the 300, 5%, or 15, may be severely retarded.[53]

In fact, for various reasons, the NRC report may have underestimated the danger. It assumes, for example, that all these evacuees were out of the zone by May 6, whereas as noted, the evacuation was not completed until at least May 21. Thus many evacuees would have accumulated higher radiation doses. Also, Prypyat was a city of young people. The average age, according to the report in *Soviet Life* of February 1986, was 26. It seems likely therefore that the figure of 300 pregnant women was too low, and in fact as will be shown, the Soviets claim to have monitored more than 2,000 babies born to evacuees from the 30-kilometer zone. One would have expected therefore that a figure of more than 100 retarded children would have been born as a result of Chernobyl (5% of 2,000, with additions as a result of the delayed evacuation).

Little information on the subject came from the Soviet side during the first year after the accident, ostensibly because the newborns were being carefully examined at the Kiev Center and other places. Subject to examination were women who had been pregnant and were subsequently evacuated although, as Zamenhof pointed out, there would also have been women who did not know they were pregnant and were not taken out of the high radiation zone immediately. Romanenko commented in March 1987—contrary to the prognostications of Gale and

Zamenhof—that more than 1,500 pregnant women who lived in the Prypyat and Chernobyl areas, as well as in adjacent villages, had given birth to healthy children. He stated that thorough examinations of the newborn babies had established that "none of them" display any deviations whatsoever from normal levels of health.[54]

Romanenko's comment was surprising news to many. It revealed, however, that the women in question had not been discouraged from continuing their pregnancies. The reason for such a policy was described by Gennadii Lazyuk, Director of the Minsk affiliate of the Scientific Research Institute of Medical Genetics of the Academy of Medical Sciences, USSR. He noted that the doses of radiation were too low to warrant recommendations of abortion (although as we have seen, such abortions were performed). A study of the condition of the fetuses during spontaneous abortions had reportedly revealed no effects of radiation on these fetuses. Lazyuk did not speculate, however, on whether the number of hereditary illnesses in Belorussia would rise as a result of Chernobyl, nor was he prepared to say what level of radiation would bring about a more difficult situation.[55]

Many women from the Chernobyl region were taken to the Ukrainian Mother and Child Protection Center in Kiev. They remained there for two-and-a-half weeks after giving birth in order to ensure that there were no changes to their health. The main concern was to establish that the radionuclide content of their milk was within the norm, in order that their children could be breastfed. The Director of the Center, Academician Yelena Lukyanova, stated that the forecasts about the effects of the accident "turned out to be worse than the reality." She maintained that the amount of iodine-131 in people's thyroid glands turned out to be within the norm, as did the penetration of the body by the isotope cesium-137, the most dangerous of the long-lasting elements thrown into the atmosphere by the Chernobyl reactor.[56]

Some 2,580 women at the Health Protection Center had been examined by April 1987. The majority of this number was checked by "traveling teams" of doctors, while 457 women and 375 newborns were seen at the Center's clinics. "There are no changes whatsoever in the status of newborns," declared Lukyanova. Her comments were superseded by a remark from Academician Mikhail Shandala, also of the Academy of Medical Sciences of the USSR. During an interview on Radio Moscow, he alluded to the interesting phenomenon that the health status of newborn children is actually better than it was among children born *before* the accident took place, an occurrence he attributes to the increased attention given to newborns from the Chernobyl region.[57]

The statement was echoed by Romanenko somewhat later, who, while acknowledging that there was a likelihood of various anomalies during delivery as a result of ionizing radiation, said that "not only did we not observe pathological changes in the course of childbirth, but we also noticed that there were fewer complications than ever before."[58] At the time that he made this statement, TASS had revealed (June 2, 1987) that over 2,000 babies born to mothers from the Chernobyl area had been examined. The question arises whether one would be unable to detect at least some anomalies in 2,000 newborns anywhere in the world, i.e., whether the Soviet medical authorities were being completely frank about the post-Chernobyl medical investigations.

In fact, there had been some curious, vaguely contradictory statements from the Soviet side. A TASS correspondent at the Mother and Child Protection Center was informed in mid-April 1987 that five women evacuees had given birth there "during the past few days." "No changes were detected in them *that could be caused by the consequences of the breakdown at the station*" [author's italics].[59] Two questions arise. First, does this mean that there were some significant deviations from the norm in these children? Second, the pregnancies in question must have occurred 2-3 months after the Chernobyl accident occurred. Thus the mothers being examined did not, strictly speaking, fall into the danger category.

Turning to children, another interviewee, E. Cherstvii, Director of the Faculty of the Minsk Medical Institute in Belorussia, noted that all children from the stricken raions were examined in detail 2-3 times in the year following the accident for changes to the thyroid gland as a result of the prevalence of radioactive iodine. This could result in serious consequences, he added, but "it practically did not occur." It was true that parents were complaining that their children were getting sick more often, but their anxieties were based on emotions.[60] The interviewer, as no doubt were the readers of this interview (in a Belorussian newspaper), was already curious: the logical deduction from the above was that "some" major changes to children's thyroid glands had occurred and that there were more illnesses among children after the accident than before.

Later in the same interview, Cherstvii acknowledged that:

Reasons for concern do exist. In a small number of children, cases have been registered of negligible deviations from the norm in peripheral blood, which are not dangerous to the health. It is possible that this is not even tied to the radiation background. Currently we are trying to resolve this difficult problem.

The main problem today, he continued, as far as children were concerned, was not iodine, but cesium, but it had been removed by the body's own mechanisms in the case of 95% of the children examined. Examining the situation in reverse therefore, problems remained with the cesium element in 5% of the children under investigation one year after the disaster.

By the summer of 1987, in the same sort of fuddled language, came the first admission from the Soviet side that some deviations may have occurred with newborn babies. Lukyanova was again the spokesperson. She informed Kiev journalists that the "vast majority" of children born to evacuees had not been affected by radiation released from the Chernobyl explosion. Tests had revealed, she declared, that "most" of the newborns had not had doses of radioactive iodine and cesium in excess of the permissible norms, based on examinations of 2,000 babies born to evacuees.[61] In brief, some anomalies had been detected. In mid-August, Lukyanova appeared to contradict her earlier statement, however, when she remarked that "no physical or mental disorders" had been detected in more than 2,000 children born to mothers who lived in the danger zone at the time of the accident.[62]

Examining the medical effects of Chernobyl is perhaps the most difficult task today for any researcher without ready access to the databank that has been set up in Kiev. In contrast to many other aspects of the disaster, the Soviet side has been noncommunicative to the point of secrecy about the effects of radiation on newborns and on children. Some deviations and ill effects clearly took place, this can be deduced from the interviews cited above, but how many occurred and to what extent remains a matter for conjecture. The investigator can, however, pose a relevant question: why are the Soviet authorities so reluctant to divulge information? Is it because the figures, on brain damage, for example, would be exceptionally high?

This is not necessarily so. The more probable reason, and one that is corroborated by the rigidity shown in other aspects, is the fear of panic among the local population that is living in the shadow of radioactive fallout. The statements of Romanenko and his colleagues become more plausible if one looks at their role as "pacifiers," calming the anxiety of Ukrainian and Belorussian citizens, who remained ignorant of the danger of radioactive fallout in the first days after the accident, and are now justifiably anxious about the consequences. The Soviet government, which failed to warn them of the danger involved, is now seeking to assure them that the impact of Chernobyl will not be significant. The innate fear of radiation has thus been given a name: "radiophobia."

Radiophobia

The term radiophobia, which appears to have been coined, with reference to Chernobyl, by Leonid Ilyin, Vice-President of the Soviet Academy of Medical Sciences, denotes "fear of radiation." It developed quickly into one of the most serious aftereffects of the nuclear explosion. As early as December 1986, Gary Lee, a reporter from the *Washington Post*, quoted Romanenko's comments about the psychological effects of the accident, which was causing one evacuee to have a blood test every 10 days rather than monthly, and the development of a "psychological trauma" in some regions of the Kiev Oblast.[63] Over the next few months, the evidence suggests that these fears increased rather than subsided.

In April 1987, Anatolii Romanenko stated that as a result of radiophobia, a portion of the population had restricted its intake of dairy products, fresh vegetables and fruit, and that the loss of such important nutritional products had lowered its resistance to various diseases "and certain negative phenomena detrimental to health." He was careful to add that the resulting illnesses were not attributable directly to radiation.[64] Later in the same month, a report on Radio Kiev confirmed that a number of people were still following the advice given by doctors in May 1986 concerning the restriction of certain foodstuffs. They were applying the same rigid diet to their children.[65]

By late April 1987, on the first anniversary of the accident, the government daily *Izvestiya* reported that a certain portion of the population was "consumed by radiophobia," which was having a more significant impact than actual radiation in the villages around the reactor. These people were afraid to leave their homes, declared the report, and refuse to eat meat, milk and fresh vegetables. Yet there was no cause for such panic: "The food products that reach our tables today are practically indistinguishable from before the accident."[66]

The radiologist Aleksei Povolyayev of the USSR Agroprom declared that such fears were making mountains out of molehills. In the Chernobyl area, he noted in July 1987, some people are so fearful that they are making other people frightened too. Every apple brought from Ukraine is regarded with suspicion, he stated.[67]

Radiophobia was not restricted to the Chernobyl area, or even to Ukraine and Belorussia, however. A report in the Bulgarian press in May 1987 referred to similar concerns in that country about the fallout from Chernobyl. "Alarm and concern" were being registered, and some Bulgarians had carried out personal checks on the food products entering their homes and had found "high radiation levels." These reports, stated the newspaper, amounted to no more than fairy tales, because the people in question were not qualified to carry out such tests on food themselves. Rather the tests should, in its view, have been left to the laboratory technicians. But some specialists also had looked at "individual results" from analyses and "made unnecessary recommendations to government agencies." This, in turn, had led to an increase in tension about contaminated food in Bulgaria. Evidently, the Bulgarian journalists' union weekly *Pogled* had published some of these recommendations, including those from "ordinary readers."[68]

The Bulgarian situation may have been brought about by the more stringent norms on irradiated food in that country. But it illustrates that psychosomatic illnesses and anxiety were not limited to the Chernobyl area. In the summer of 1987, the most detailed and thorough analysis of radiophobia to date was provided by Ilyin,[69] in an attempt to allay finally the fears of the population.

Ilyin stated that radiophobia is not an illness, but a condition, namely the fear of the biological influence of radiation. He described it as an "extraordinary complex" that was very hard to resolve. Those people suffering from radiophobia, he added, do not believe anyone or anything, and connect the most trivial ailments with the effect of radioactive substances. As a result, the common illness actually becomes something more complicated and the person "does himself a terrible service." One can deduce from his remarks also that once the illness has become more complicated, it is all the more logical for the patient to feel that he must indeed be suffering from radiation sickness—in fact the line between direct and indirect radiation sickness becomes very fine indeed because radiation is the ultimate cause of the complications.

But why did radiophobia appear in the first place? Ilyin responded that a very serious reason was ignorance of the problem, even at the level of some medical specialists, let alone at the level of the general population. "The level of general radiation literacy is low." He cited the fact that immediately after Chernobyl, certain Kievans rushed out to take neat iodine, in their ignorance, and that some of them had become ill in consequence. Even before Chernobyl, such misfortunes had occurred, said Ilyin. Thus a group of women irradiated in Japan by the 1945 atomic bombings developed breast cancer. In their anxiety about this observation, they subjected themselves to continual x-rays, which only accentuated their condition. Once again, a lack of information had brought on radiophobia, which is exactly what had occurred after Chernobyl.

Ilyin's theories in this regard seem perfectly plausible. However, when he applies practical examples to illustrate them, he falls into the same sort of niche as Romanenko, with his constant denial of any ill effects among newborn babies. In this respect, a vicious circle is created. Thus, as will be seen below, Ilyin distorts some of the events of the Chernobyl affair in order to demonstrate that the population was in less danger than was actually the case. When his remarks are contradicted by forthright journalists, such as Andrei Pralnikov of *Moscow News*, for example, the average reader may then recognize that Ilyin is not being strictly accurate and then refuse to believe much of his generally sound medical advice. The problem becomes compounded, but it is a problem that began with the silence about the accident in the first hours after its occurrence. In this lay the seeds of radiophobia and the subsequent illnesses in the villages around the reactor.

Ilyin, for example, attacked those writers—they were described as "they"—for stating that the evacuation of Prypyat was delayed. "One can imagine the mood of the Prypyatites!" According to Ilyin, however, they were only evacuated to assuage any anxieties about the situation. "Prypyat was evacuated in a condition that did not threaten the health of the people." Almost at the same time that Ilyin was making this comment, Viktor Bryukhanov, the former plant Director at Chernobyl, was receiving a 10-year sentence for the delay in evacuating Prypyat and for releasing figures on the radiation level that were "dozens" of times lower than reality, as Pralnikov was writing. At this same time, the local party workers, as noted above, were screaming for vengeance against those leaders who had subjected their families to peril by not having them removed from the city at the earliest possible opportunity. Ilyin's statement is without foundation.

Ilyin next denied official Soviet reports that the film director Volodymyr Shevchenko died of radiation sickness. "Certain cinematographers," he grumbled, "connect the death...of V.N. Shevchenko...with the accident at the atomic energy station." His ailment, in Ilyin's view, began before Chernobyl. He denied that it appeared as a result of radiation and declared that such allegations are "inhumane" because they implant "abnormal doubts" (i.e., radiophobia). He then proceeded to lecture on how information about such an accident should be presented to the public:

> During tragic events, not only is the truth important, but also how it is presented. Also of significance is the tone of the conversation and even intonations. Simplification and superficiality,

lightheartedness, let alone inaccuracy, are intolerable. They breed a distrust of information, and helpful recommendations do not find the desired application.[70]

How does one treat radiophobia? In Ilyin's view, the first solution is to remove its causes, to educate the population about the actual effects of the radiation. But these lectures should not be conducted by "dilettantes," since such people only aggravate the situation. One must listen to specialists and doctors rather than rumor-mongers. He referred to some students, who went around the local villages, exaggerating the effects of radiation in an attempt to persuade the inhabitants to purchase dosimeters. Having bought the instrument, the villager was incapable of using it properly, again with unfortunate results.

Another problem (which will be examined in more detail below) was that some specialists had allegedly provided unwarranted computations about the number of future cancer victims as a result of Chernobyl. In turn, this only aggravated radiophobia. One such person was Dr. Gale, whom, said Ilyin "is a major specialist, but not in the area of radiation medicine. He was mistaken in his calculations." Other people were indulging in gossip about cancer victims already appearing, even though on average the cancer will take 20-23 years to appear, according to Ilyin. The sequential conclusion from this line of thinking, one should note, is that any specialist who from the best of motives, arrives at a rate of future incidence of cancers that is significantly higher than the official Soviet figure risks denunciation for causing panic. One can appreciate Dr. Ilyin's motives, therefore, but his statements about actual events automatically become suspect.

Moreover, his reasons for the development of radiophobia are somewhat one-sided. In some areas around the reactor, the local food was contaminated, and supplies had to be brought in from the outside. In one instance, however, concerning the Krasnopolsk district of Belorussia, outside supplies of food were nonexistent. There were no refrigerators in the region, milk brought in from the outside by cart was "sour by the time it arrives." The local population was simply unable to acquire supplies of meat, milk and other dairy products.[71] In short, radiophobia may not always have been the problem. Some local residents had been physically cut off from supplies of uncontaminated food, and thus became ill as their normal diet was curtailed drastically.

Finally, it is worth reiterating: radiophobia arose from a lack of accurate information about the radiation situation after the accident.

The damage was done at the outset. Ilyin and others were thus given the task of rectifying a situation that was not of their making (although Romanenko, as the Ukrainian Minister of Health, must bear some of the responsibility for failing to inform the population about the state of affairs for 10 days). Whether anyone actually died of radiophobia is not known. But its seriousness is not in question. After the first, dramatic and tragic Chernobyl deaths, and before the inevitable cancer that will consume future victims in the thousands, came radiophobia. It existed from the first days after the accident and occupied a prime place in the Ukrainian press and on the radio stations at least until July 1987. Almost equally important, it led the Soviet authorities continually to underemphasize the impact of the disaster in order to allay the fears of the population. In the longterm, this may prove to have been more damaging than the disease itself.

Future Casualties

The question of the ultimate casualty figure from the Chernobyl disaster has become a matter of the utmost controversy. The degree of variation is so immense—from 200 to 500,000 longterm deaths—that there may be grounds for considering such conjecture an exercise in futility. Nevertheless, although this author is neither capable of nor willing to make such prognostications, he perceives a certain value in the process because it is perhaps the only way in which we can know how many actually suffered from the nuclear fallout. As has been pointed out frequently, the high rates of cancer deaths in industrialized societies (perhaps 1 in 4 deaths) mean that it will be impossible in the future to perceive the impact of Chernobyl. Even if one takes the highest figure offered of 500,000, these victims will simply be swallowed up in the overall figure. Since the Soviets may not be entirely forthcoming about the fate of the 100,000 or so individuals being monitored at the Kiev Center, we are dependent upon speculations about the final death tally.

At the outset, it should be borne in mind that the Soviet total has decreased consistently over time. Further, Soviet theoreticians have frequently provided data not from an objective examination of the fallout, but rather for the sole purpose of discrediting Western estimates of the future death and illness toll. Finally, as already noted, the cleanup workers appear to have been omitted constantly from the estimates. The initial Soviet report to Vienna, upon which the early estimates of casualties were elaborated, never mentioned their presence.

We can begin with Dr. Robert Gale. In his appearance on CBC Television (Toronto) in June 1986, Dr. Gale attacked those scientists who were speculating about future Chernobyl-induced cancers. Eventually, however, he appears to have been prompted (perhaps because he had more information at his disposal than most of those making the predictions) to provide his own figures, and it is these figures that have been at the heart of the Soviet campaign to underplay the impact of the disaster.

In September 1986, in Las Vegas, Nevada, Dr. Gale estimated that 2,500-25,000 Soviet citizens would die from the Chernobyl accident over the next 70 years.[72] Like other experts, he based his deductions on the figures on radiation fallout per person in the fallout area from the figures provided by the Soviet side to the IAEA in August 1986. Two months later, in Milan, he maintained that there would be 5,000-75,000 cancer deaths worldwide, while noting that he believed that the real figure would be closer to 10,000. The most common ailment, he stated, would be leukemia, cases of which would begin 2 years after the event (May 1988) and the largest number of such cases after 7 years (May 1993). Later, other types of cancer, such as breast, lung, gastroenteric and thyroid, would make their appearance.[73]

In June 1987, at a news conference in Bonn, Gale cited "experts" as estimating that the Chernobyl accident would cause up to 60,000 additional cancer deaths worldwide over a 50-year period. He also foresaw up to 1,000 birth defects and up to 5,000 "genetic abnormalities." He stressed that only 40% of the defects would occur in the USSR. However, the main birth defect—severe mental retardation—cannot be confirmed until the child has reached the age of two.[74] The U.S. doctor reiterated his upper limit of 75,000 extra cancer deaths worldwide over the next 50 years in July 1987, but reduced his lower limit to 2,500.[75] During a telephone conversation with the author in the Fall of 1987, Dr. Gale stated that he basically agreed with the "circa 30,000" excess cancer fatalities figure advanced by the U.S. Department of Energy.

In the Soviet Union therefore, Dr. Gale's estimates are for excess cancer deaths of between 2,500 and 30,000. Most other Western estimates have been in the same area. For example, the report of the U.S. Nuclear Regulatory Commission estimated that among the 135,000 evacuees, one could expect about 320 excess cancers that will be caused by the external radiation, although about 16,000 of these evacuees can be expected to die of cancer during the "normal" course of life. As for the 75 million exposed to additional radiation fallout in the European part of the Soviet Union, allowing for Soviet overestimation of the dose through the food channel by an order of magnitude, the NRC's conclusion is that

> one estimates a total collective dose of about 5×10^7 person-rem. Assuming a risk factor of 2×10^{-4}/rem, about 10,000 fatal cancers (plus a comparable number of nonfatal cancers) would be projected over the next 70 years. Mitigation measures will reduce the collective dose, but final consideration of the enhanced cesium uptake by crops in the Poles'ye region as well as inclusion of neglected sources of exposure (e.g., from contaminated crops in other parts of the Soviet Union, from uptake of Sr-90 [strontium-90], and from inhalation of resuspended radioactivity) may substantially increase the final estimate.[76]

The report by the U.S. Department of Energy that was published in the summer of 1987 raised the number of expected cancers, although its figures remained within the limits projected by the NRC and Dr. Gale. Authored by Marvin Goldman, Professor of Radiation Biology at the University of California, the report predicted 12,000 fatal cancers in the Soviet Union, and a further 21,000 in Europe over the next 50 years. Goldman and his colleagues combined the information provided by the Soviet side with the radiation measurements outside the USSR. Perhaps the main difference from the earlier NRC study was the finding that considerably more cesium-137 had been emitted from the exploded reactor than first thought. Otherwise the methods used to make the predictions were similar to those used before.[77]

As final examples from the West (before examining Soviet estimates of the impact of the accident), the report of the Central Intelligence and Defense Intelligence Agencies (CIA/DIA), released in March 1987, stated that there would be at least 500 cancer deaths among the 135,000 evacuees, as compared to the NRC estimate of 320 cancer deaths in this same group.[78] As noted above, the NRC figure assumed the speedy completion of evacuation, whereas in reality it dragged on for almost a month.

The Soviet authorities have not accepted Western estimates of future cancer casualties as valid. On May 6, 1987, TASS issued revised figures on the irradiation of people within the 30-kilometer zone, which it estimated at an average of between 5 and 10 rems per person (rather than the original estimate of about 12, i.e., about half of the original total). This was much lower than originally thought, and there were some doubts expressed by Western representatives in Moscow as to whether such low figures could have caused the extensive environmental damage to the zone around the reactor.[79] Romanenko commented in this same month that the extent of irradiation in the Chernobyl zone was much lower than the "accident rates" of both national and international radiation protection organizations. He maintained that outside the 30-kilometer zone, and in the Ukrainian SSR generally, the radiation doses received were comparable to the natural radiation background.[80]

Ilyin again emerged as the major Soviet spokesperson. Interviewed in a Ukrainian newspaper in mid-June 1987, he announced that Soviet scholars had been working on the subject of excess cancers with experts from the IAEA. Their conclusion was that any increase in cancers resulting from the irradiation of persons—and by this he signified only those who worked at the station immediately after the explosion—could be measured only in hundredths of a percent. Having criticized Gale once again for making predictions of future casualties, he stated categorically that "any discussion of an increase in cancer illnesses...is nonsense."[81]

On this subject, however, the Soviets mounted a major assault on the Western study groups. One attack was provided in a special issue of the *Voyennyi vestnik*, by a leader of the chemical troops of the USSR, Colonel-General Vladimir Pikalov. His view was that the evaluation of the radiation consequences for those people subjected to radiation contamination from the accident has revealed that the "value of the external dose of gamma and beta radiation" for the year 1986 was lower than the annual limit established by the norms for radiation safety in 1976. Consequently, he stated, the increase in the number of additional cancer fatalities will represent less than 0.05% of the level of fatality of spontaneous cancer. As for the cesium-137 element, examinations had shown that the likelihood of extra cancer deaths from radioactive cesium did not exceed 0.4% of the natural death from malignant neoplasms. Even these calculations, according to Pikalov, concerned only the regions adjacent to the reactor. For other areas there are no grounds for expecting any additional cancers, thus "Western speculations" of 50,000 excess deaths "must be dismissed as totally groundless."[82]

Writing in *Izvestiya*, I. Likhtyarev, Head of the Physics-Mathematics section of the Center for Radiological Medicine in Kiev, stated that:

The calculations are convincing: in 70 years of life, the dose to the population of the accident regions can be no higher than 10-15% of the background, that is not included in the accident background. It is clear to any unbiased person that these levels of irradiation do not call for realistic anxiety.[83]

To these comments can be added others made in September and October 1987, during which time the Soviet nuclear and health authorities made another presentation to the IAEA in Vienna, and once again strongly criticized "scaremongers" in the West who had spoken about the possibility of "1 million tumors" as a result of Chernobyl. Romanenko, interviewed by TASS on September 11, 1987, displayed his optimism; "We do not expect the accident in Chernobyl to cause any unpleasant genetic consequences." He referred to the radiation doses of the people evacuated from the danger zone and others in adjacent zones as ranging from 0.12 to 0.74 of the biological equivalent of the roentgen. Thus, he maintained, if the level of radiation absorption does not increase, then it will pose no dangers even 50 years hence.

Speaking at a press conference at the time of the 1987 Vienna meeting, Ilyin again denounced "baseless" Western assessments of malignant tumors, and moreover, refused to provide any figures of his own, on the grounds that such predictions could only be theoretical—which, of course, is true. Nor did a paper he authored jointly at the conference provide any enlightenment on this same question.[84] On October 15, 1987, Vasilii Politchuk, a Ukrainian delegate to the United Nations, said that there was no reason to be concerned about either short or long-term medical or biological consequences of the Chernobyl disaster. The latter, in his view, was effectively over.[85]

None of the above Soviet reports and comments signify that the Soviets believe there will be *no* future Chernobyl casualties, although a brief perusal of the press items might lead any unenlightened reader to this conclusion. Rather Ilyin and his colleagues have adopted the line of the lowest possible radiation fallout and the least number of victims in the future. As noted, the fallout of cesium, iodine and other isotopes, a topic on which the information provided is spotty, has been reduced by the Soviet authorities. Their conclusion is that Chernobyl, while a terrible accident, will not be quite as bad as originally suspected.

Why has this view been adopted? In the main there are three principal reasons:

1) Radiophobia among the local population.

2) A growing sentiment against nuclear power in parts of the Soviet Union that can only be accentuated by news of high future casualties.

3) The development of a "myth of Chernobyl."

As the first reason has been dealt with above, and the second will be discussed in detail later in this book, only the third reason requires an explanation. The history of the Soviet state, by and large, is not endowed with too many outstanding successes. The major event, and one about which any visitor to the Soviet Union will have been reminded almost everywhere he cares to go, is the victory over Nazi Germany in World War II. It seems plausible that in addition to the enormous sacrifices of the Soviet Union in this war, the latter also supplied the Stalin regime with a historical legitimism that it lacked hitherto. A country that had been isolated and subsequently regarded as a short-term anathema by many Western politicians had finally come of age.

In the history of the Chernobyl accident, images of World War II resurfaced: a fight against seemingly insuperable odds; selfless heroism; ultimate triumph and victory. Chernobyl *was* a victory, if, as the Soviets state, only 31 deaths resulted from the world's worst nuclear accident. But if there were more casualties than stated, and if in the longterm there will be thousands more deaths, the victory becomes frayed, something akin to an illusion on the part of the Soviet leadership. After the initial Soviet silence about the accident, a plethora of articles clarified and ultimately mythologized the entire event. There were heroes and villains, good and bad, and the party was ultimately the decisive factor—in the Soviet view.

Thus when an objective observer like Dr. Gale provides an "upper limit" of 75,000 cancer victims, he is unwittingly taking a stance against what has become the official party line on the Chernobyl accident. As such, even as a "friend of the Soviet Union," he had to be verbally shot down. Moreover, when a film director who was rash enough to film in the area immediately after the accident dies as a result, other reasons are found for his death. It is ironic that in an era of openness, Chernobyl may have been both the pioneer of glasnost under General Secretary Mikhail Gorbachev and then subsequently its first casualty. The vast majority of the immense problems that have followed the event, and which continue to plague the Chernobyl region today, are a result of the official view of the disaster and attempt to minimize its consequences in official statements and publications.

Finally, it should be emphasized that while there have been individual Western speculations of an extremely high number of future cancer deaths from Chernobyl, the reports, such as that of the NRC, are notably sober and perhaps even conservative in analyzing the likely impact of the accident. The NRC report puts the matter into perspective:

> Although the number of excess fatal cancers predicted on the basis of the Soviet report is very large, these will be widely distributed over a population of 75 million people and over decades....Correspondingly, the risk to an average individual in the population owing to the accident is relatively small...the estimated individual dose from external and internal pathways would be about 0.67 rad, roughly equivalent to the dose received from background radiation over a period of 7 years. Based again on a risk factor of 2×10^{-4}/rad, this dose would give an estimated lifetime risk of 0.013%, which is only about 0.1% of the stated Soviet baseline risk of fatal cancer (12-13%).[86]

In short, therefore, the danger to those in the 30-kilometer zone, and especially in the vicinity of the reactor is high, but for the average citizen of the European USSR, the chances of future fatal cancer as a result of Chernobyl are too low to be measured.

2 The Environmental Impact

> [The inhabitants of the] USSR do not know their future: neither the radioactivity level of their locality nor the extent to which the products they buy or grow are radioactively contaminated....Everything continues to be decided by those who conducted the Chornobyl experiment and now supervise the radioactivity and their secretly held measurements.
>
> (*Samizdat* document, March 1987)

Western scientists, using data from both Soviet and non-Soviet sources, have estimated the future cancer rates as a result of the disaster. They are on less secure ground when trying to assess the environmental impact of the accident. As the above quotation reveals, some Soviet citizens themselves remain uncertain about the consequences. The situation was complicated by the unevenness of the fallout and the contradictory statements issued by the authorities regarding the need for precautions. This chapter will also examine the question of the water supply in the Chernobyl area, but it should be stated at the outset that no figures on water contamination have been made available.

The first Soviet report to the IAEA, of August 1986, noted that there were two major periods of radioactive fallout: April 26-27; and May 2-6, when the reactor heated up for a second time for reasons unexplained, but probably to do with the dumping of 5,000 tons of lead, dolomite and sand through the hole in the roof by helicopter to quench the graphite fire. The radioactive plume reportedly reached a height of 1,200 meters on April 27, and 3.5% of the fission materials in the reactor core were released into the atmosphere. Both the 1986 and 1987 reports to the IAEA in Vienna cite a maximum radiation level in the vicinity of the reactor (5-10 kilometers) of 1 rem/hour. The ground became severely contaminated to the west, north-west and north-east of the Chernobyl plant, while the plume itself affected the Ukrainian, Belorussian and Russian republics, before moving northward across the Baltic republics and Scandinavia.[1]

The Chernobyl area is not a major agricultural region. It consists largely of marshlands and peat bogs. The agriculture that does exist is mainly that of dairy farming and hardy crops such as potatoes. However, it is connected directly with the main river systems of the Ukrainian republic: the Prypyat and Uzh Rivers run into the Kiev Reservoir which, via the Dnieper River, provides the water supply for the 2.4 million residents of the city of Kiev. The Dnieper then flows southward dividing Ukraine into its Right and Left Banks. Through irrigation and diversions of its main flow, it services valuable agricultural land in Ukraine, which is not well endowed with water. To the north-east, the Desna River runs southward toward Kiev from Chernihiv Oblast. The lakes in the region have been used widely for fishing, even including the cooling pond of the Chernobyl nuclear plant, the location of a fish farm. In brief, then, the environmental impact of the accident is considerably wider than it might appear initially.

Environmental Control

After the accident, the Soviet authorities set up a system of radiation control in various parts of the Ukrainian SSR. Altogether, 200 stations, including 16 in the city of Kiev, were created to monitor the atmospheric air, soil and water for the rate of gamma background radiation. The main areas were said to be Kiev, Zhytomyr, Chernihiv and Cherkassy Oblasts. Some 70,000 cubic meters of air were passed daily through special filters and the filter material was changed each morning at 9am. The old material was put into a packet and burned in the laboratory inside a crucible in a special furnace. After a certain time period had elapsed, the ashes of each filter were then analyzed by instruments. Tests of soil were also conducted in open fields that had not been used for any sort of farmwork after the accident. These were enveloped in polythene and tested in a "special cast-iron protective building" in order to prevent any effects from external forces.[2]

The conclusions from such tests have been mixed. The Chief Engineer from the Dosimetric Control Service in the Kiev region, Dmytro Voskresnykov, noted as late as July 1987 that there were still contaminated areas in the zone.[3] Strict rules remained in force. These were described in the reports of Western journalists who visited the Chernobyl area in the summer of 1987. Thus Thom Shanker of the *Chicago Tribune* commented that in the city of Chernobyl—some 22 kilometers from the damaged reactor—the foreign journalists had to acknowledge in writing that they were aware of the special risks of travel in the area. Bus windows had to be closed and it was prohibited to walk off the roads or into the forests. It was also not permitted to eat and drink outside the bus and before entering a building in the city, one had to use the hoses placed at the entrance to wash down one's shoes in order not to bring radioactive dust into the buildings.[4]

As noted, the city of Kiev was provided with 16 monitoring points. As it possesses the largest population in the area and the main market center for the distribution of agricultural goods, Kiev became the fulchrum of the process of environmental control. It is somewhat ironic that the city which organized its May-Day parade almost in defiance of the lethal disaster should subsequently become one of the safer places in the region as a result of careful monitoring. However, the devices for thorough inspections and above all the necessary personnel came to Kiev after the accident. The more northern areas closer to the reactor were neglected by comparison.

According to Soviet figures, the highest radiation levels in the city of Kiev were 0.5-0.8 millirems/hour, which were reached early in May 1986.[5] Although the Soviets have always claimed that this level was never dangerous, these figures represented an increase over the average radiation norm of 160-300 times, an enormous, if short-lived rise. By May and June 1987, the levels had fallen almost to the normal background rate at 0.01-0.02 millirems per hour. O.V. Serebryakov, the head of the Ukrainian Center for Radiation Control, remarked that the levels had actually stabilized in the Fall of 1986,[6] hence it is hard to avoid the conclusion that Kievans were actually in the greatest danger either at the time of or immediately following the May-Day parades, presided over by Ukrainian First Party Secretary, Volodymyr Shcherbytsky.

In April 1987, a year after the accident, M. Mukharsky, Deputy

Director of the Main Sanitary-Epidemiological Department of the Ukrainian Ministry of Health, effectively nullified the original health warning issued to the citizens of Kiev on May 5, 1986. Mukharsky noted that "The radiation situation in Kiev does not necessitate any restrictions whatsoever in the lifestyle or behavior of its residents." Paraphrasing Romanenko's 1986 warning, but now in the positive rather than negative mode, he announced that Kievans could relax in the green zone, bathe in the Dnieper and fish from the same river, all of which indicated that they had been restricted from doing such things in the very recent past.[7]

There followed one of those curious post-Chernobyl enigmas that have often negated any kind of objective analysis. On July 9, 1987, on its World Service, Radio Moscow stated that the Kiev City Council had decided to end its controls on radioactive products at city markets. The Director of the Institute of Environmental Physiology (whether this was a republican or all-Union institute was not made clear), Mikhail Shandala, informed the Radio that since checks had not registered any radioactivity in food products coming on to the market "for a long time," the controls were to be lifted. The news was broadcasted twice during this same afternoon by Radio Moscow, but as far as is known, it was never transmitted to a domestic audience.

Just one week later, Leonid Ilyin was interviewed by a Ukrainian newspaper correspondent and the conversation that ensued also focused on the radiation situation in the city of Kiev. The correspondent stated that "in Kiev, radiation control over food products is still in operation," and that people in the city were still putting wet rags in front of their doors and rinsing areas that they had not rinsed before. Ilyin adopted his now familiar role as a "calmer of anxieties," stating that it was always wiser to be careful and that one should keep one's house clean and in order. He reiterated Mukharsky's comment about swimming in the Dnieper, adding the enjoyment of "fresh fruits and vegetables" and sun-tanning to the list of advisable recreational pursuits. But he also revealed that control over food products at the Kiev market would last "for more than one more year," thereby contradicting the Radio Moscow broadcast of a week earlier.[8]

In early June 1987, Anatolii Romanenko, Minister of Health Protection of the Ukrainian SSR, and Director of the All-Union Center for Radiation Medicine in Kiev, had also stated that all produce in the stores and on the markets in Kiev (and neighboring regions) was undergoing rigorous monitoring. Such control, he continued, "will naturally continue in 1987 and the following years" because radiation will begin to enter the crops through the root systems of plants (see below).[9]

How does one explain such a divergence of interpretation? Either the controls were rescinded or they were continued. Since Romanenko is the main Ukrainian health authority, and Ilyin has emerged as the main Soviet spokesperson on Chernobyl both at home and internationally, one is inclined to believe their statements that the controls over radiation contamination were still in place. This suggests that the Radio Moscow broadcast was intended to assure foreign listeners that the situation in Kiev had been "normalized," even to the extent of falsifying the information. On the other hand, the broadcast could have been issued in error, although the fact that it was reissued on the same afternoon tends to mitigate against this possibility.

Kiev itself was on the periphery of radiation fallout. The extent of the protective measures adopted there is a good indicator of the scale of the problem in the 30-kilometer zone itself. In fact, the level of external radioactivity 15 days after the accident, 60 kilometers from the damaged reactor remained at 1,000 times higher than the normal background, based on the Soviet report to the IAEA of August 1986.[10] Thus even areas that were initially considered safe—the Poliske region is the prime example—were now suffering from high levels of additional radiation. Under these circumstances, the process of monitoring the environment became a protracted, complicated and highly uncertain affair.

Rivers, Lakes and Forests

Immediately after the accident, the Soviet authorities began to take water samples from the rivers Teteriv, Prypyat, Irpen, Desna and Dnieper (but curiously enough, not the Uzh). The first report to Vienna stated that the highest concentrations of iodine-131 were observed in the Kiev Reservoir on May 3, 1986, and that the spatial distribution of the radioisotopes was very uneven.[11] Time has shown, however, that the principal danger to the water supply is not iodine-131, but cesium-137. As far afield as Lake Lugano in Switzerland, for example, governments have banned fishing on lakes as a result of the cesium fallout from the Chernobyl No. 4 reactor. Thus a heavy rainfall deposited "unacceptably high levels" of cesium into Lugano in late August 1986, and the Swiss government banned all sales or imports of fish caught after August 1, 1986, while noting that the ban would be in force for at least 3-4 months.[12]

These restrictions also applied to other areas. In January 1987, Ian Waddington, the Director of the Environmental Health Service of the European Regional Office of the World Health Organization, pointed out that "a real long-term problem" after Chernobyl was the contamination of lakes. The main difficulty was to remove cesium collected in sediments from the lakes. Waddington said that there would probably be longterm restrictions on fishing and fish consumption in Finland, Norway and central Sweden.[13] Again, the implication is that closer to the exploded reactor, the situation could only be worse.

In June 1987, when foreign journalists were permitted to visit the Chernobyl area, they were addressed by the station Director Mikhail Umanets and other officials. The Soviets commented that all the long-lived heavy nuclides that were emitted from the exploded reactor and found their way into the water supply have settled at the bottom of lakes and rivers and "are now in an insoluble phase." Over the next 800 to 1,000 years, it was reported, these nuclides will not dissolve.[14] On the same topic, Radio Kiev, in a broadcast of June 16, was somewhat more assuring. Acknowledging that in certain locations, radioactive traces had landed on the surface of open reservoirs and eventually settled on the bottom, it stated that these sediments did not affect the quality of the water, which was reportedly within acceptable norms not only in the area around the city of Prypyat, but even in the cooling pond of the nuclear plant.

Two months earlier, in April 1987, a Ukrainian correspondent interviewed L.M. Khitrovy, a radiochemist, who directs the laboratory of the "V.I. Vernadsky" Institute of Geochemistry and Analytical Chemistry of the Academy of Sciences of the USSR. Khitrovy has headed a group of specialists that has been investigating the influence of the disaster on the surrounding reservoirs since May 1986. Khitrovy revealed that a map of the Kiev Reservoir, "which accepts all the minor rivers of the 30-kilometer zone," had been drawn up each month since the accident. He also stated that the Dnieper and all its cascades had been examined twice, from the mouth of the Prypyat all the way to Kherson, and that there were no deviations "from the medical norms." But surely, the correspondent asked, the situation in the reservoir can change in a day, or even an hour? Khitrovy responded that "operative analyses" were also conducted on a minute-by-minute basis at various points on the rivers and reservoirs.[15] Nevertheless, the correspondent's point—that precise monitoring was almost impossible—was well taken.

The first priority of the Soviet authorities was the Dnieper River, Ukraine's most important water supplier and the main source of the republic's agricultural wealth. Khitrovy hinted that between May and October 1986, the river may have been contaminated in the Kiev area. Early in July, the water supply from the river to the city of Kiev was abruptly cut off and an alternative supply from the Desna River was harnessed (although the authorities may have had few alternatives open to them, the Desna may not have been entirely safe either in that it was also in the direct path of radiation fallout).[16] No explanation was given as to why there was a two-month delay in implementing this basic safety measure. The Soviet side has never admitted openly that the Dnieper was contaminated, but the lengths to which they went to provide the new water source from the Desna indicates that this was probably the case.

Not until December 1986 were supplies of water from the Dnieper used again. The time period is somewhat longer than would have been needed for a routine inspection, *had that inspection proved that fears of contamination were groundless.* More recent interviews with Soviet officials have tended to play down the issue. Several have claimed that the Dnieper was never contaminated at a level that could have caused danger to the population either using the water supply or swimming in the river (the latter would have been more dangerous because of the raising of sediment from the bottom of the river). For example, O.V. Serebryakov (cited earlier), stated that water specimens had been analyzed weekly for isotopic content from the Dnieper River and those rivers that empty into the Kiev Reservoir. In the reservoir and the Dnieper River, he remarked, "severe changes in beta-activity were not observed, even during the fallout of sediment."[17] The key word in this quotation is the adjective "severe." It implies that changes did occur, but were not serious enough to cause problems. Yet the water supply was subsequently cut off. Again one is faced with an ostensible contradiction because the Soviet authorities are not willing to admit that a major problem existed.

The radioactive materials threatened to penetrate the Ukrainian river tributaries in the first weeks after the accident. By October 1986, with the use of a soil-machine, an underwater dam had been constructed, which was 450 meters in length. At the front of it was created a groove of 100 meters in width and 16 meters deep. Its purpose was to catch radionuclides entering the Kiev Reservoir from the tributaries of the Prypyat River. A second silt trap was created in front of the dam at the Kiev hydroelectric station, made of crushed stone and again with a wide groove at its front part. Water-retention structures were constructed on the rivers Sakhan, Veresnya, Berezhest, Radynka, Braginka,

Nesvich and others that empty into the Uzh and the Prypyat Rivers.[18]

A key concern in the first weeks after the disaster was the state of the nuclear power plant's cooling pond. After the accident, noted the *Pravda* correspondent, M. Odinets, radioactive substances fell into the cooling pond. This appears to contradict the statement made by Evgenii Stukin of the State Committee of Hydrometeorology and Environmental Control, who had denied that the cooling pond had been contaminated by radioactive fallout in an article of May 1987.[19] However, Valentin Sokolovsky, the Deputy Chairman of this same committee, concurred that the water in the cooling pond showed increased levels of radioactivity and required constant monitoring.[20]

As the level of that pond was 7 meters higher than that of the Prypyat River, it seemed probable that the natural reservoirs would soon become contaminated. A 30-meter wall was built into the ground at the No. 4 unit that reportedly blocked the movement of ground water toward the Prypyat River. At the entrance to the latter, a drainage screen was established. Two brigades from Baku were employed in drilling boreholes in the cooling pond.

Such measures resolved part of the problem. The next task facing the Soviet authorities was how to cleanse the radioactive water. Special tanks had been used to hold the contaminated water—clearly it could not be permitted to drain back into the water supply system—but such storage had its limitations (the same problem, incidentally, cropped up with the storing of radioactive topsoil). As the traditional cleansing methods were difficult to employ and expensive, the members of the civil defense employed in this work turned for help to the Kiev Polytechnical Institute. At the latter, scientists had been working on experimental procedures for cleansing contaminated industrial waste water by using reagents from industrial wastes. These same reagents were applied to the water contaminated from Chernobyl, directly into the tanks, apparently with remarkable success.[21]

The authorities were responding to the crisis piecemeal. As they encountered new problems, then new devices were created and employed to deal with them. The time period of the accident was a crucial factor in that it marked the end period of spring flooding. Dry and imperceptible rivers suddenly filled up with water. Among the multitude of such rivers around the nuclear power plant: "runoffs cannot be avoided, the flow is not stopped." In some places a dam was to be established, in others a dike would suffice. Each section was examined from the air by helicopter, but no sooner had solutions been found than new difficulties would present themselves.[22] In short, then, during the time it took to come up with radiation protection devices, the danger of widespread contamination of tributary rivers increased all the more. The question that is not answered by Soviet scientists is how much contamination of the water supply occurred during this process.

Frequently, the necessary technology to dig the grooves for radionuclide retention networks was simply lacking. For the most part, foreign offers of assistance were accepted, but then "implacable customs officials" refused to allow trench-digging complexes into the country until all the necessary formalities had been met. While the machines were being delayed at the border, the Soviets were working to produce their own "more powerful complexes." These were built very quickly, but nevertheless, another month had elapsed before they were ready. For some rivers, such as the Braginka, the dike required a filter that was almost 500 meters long. This took two months to construct.

Even more complicated was the building of the wall at the Chernobyl plant. Specialists expressed some doubts over whether a narrow trench dug into the sand at such a depth might not simply collapse. Eventually the problems were overcome, "an enormous monolithic curtain" blocked the flow of groundwater from the No. 4 unit to the Prypyat River. It had taken from May until September for the work to be completed. The workers labored for arduous six-hour periods wearing respirators. The shields were evidently effective, but how much contaminated water had flowed through during the period of construction?

In addition to lakes and rivers, forests were also contaminated by the radioactive fallout. Close to Prypyat, a forest with tall pine trees absorbed the highest fallout, turning into a rust-color as a result. It became known locally as the "Ruddy Forest," and was frequently mentioned by those who first went to visit the Chernobyl plant after the accident. Ian Waddington, cited above, noted that whereas deciduous trees with cesium-contaminated leaves shed these same leaves in the Fall of 1986, fir trees shed their leaves only every four years. There were thus two main areas of concern: forest fires, which could quickly spread radioactive cesium; and falling pine needles that might contaminate the forest vegetation in the vicinity.[23]

Even in October 1986, Romanenko stated that the falling leaves from the deciduous trees had raised the radiation level of other objects in the environment. The leaves, he said, were to be removed, but would definitely not be burned.[24] Over the course of more than a year, part of the redwood pine forest near Prypyat was chopped down by construction machines. The dead wood was then covered with sand and treated with a chemical preservative. The "hot" parts of the trees were placed in containers and then buried in concrete. By June 1987, noted a Western reporter, about half of this forest had actually been chopped down. A Soviet source, however, implied that the entire task had been completed a month earlier as it referred to the forest near Prypyat where "tall pines...were growing only recently."[25]

The Spring Floods of 1987

The spring floods of 1987 constituted the clearest of the tasks being encountered by the authorities in the wake of Chernobyl. As noted, even in the summer of 1986, steps were being taken to prepare for the perennial problem in many areas of the European USSR: the melting of snow followed by extensive flooding. The problem had been exacerbated by an exceptionally early winter that had begun with a snowfall on September 27, 1986 and continued to be unusually cold. During an interview with a U.S. Senate panel headed by Senator Edward Kennedy in January 1987, Evgenii Velikhov had focused on spring flooding, commenting that the spring rains in the marshlands around the Chernobyl plant usually cause an increase of up to 10 times in the water-flow rates, but he was confident that the dams and bypass canals built in preparation for the floods would prove adequate for the task.[26]

The main threat posed by the high waters was outlined by Ian Waddington in January 1987. He pointed out that spring flooding could bring the contaminated groundwater to the surface and the latter might then flow into the river systems, particularly the Dnieper River. If the walls and embankments proved inadequate to hold the water, then a significant release of radioactivity might result.[27] Chernobyl, then, had increased the concern of the authorities for an annual phenomenon. In Kiev Oblast, the task of containing the spring floods was supervised by a Commission on Fighting Natural Phenomena, headed by the First Deputy Chairman of the Kiev Oblast Executive Committee, M.S. Stepanenko. In early March 1987, Stepanenko outlined the main tasks facing the Commission.

Stepanenko emphasized that the Commission based its work on the February 1987 prognosis of the Ukrainian Hydrometeorological Center. The latter had pointed out that measurements of snow at the beginning of February had revealed that the surplus of water in the snow was 2-2.5 times higher than the norm. "This cannot fail to prick up your ears." The Commission's main anxieties concerned the Dnieper River, which extends for 246 kilometers within Kiev Oblast, and its main tributaries such as the Prypyat, the Teteriv, and the Kiev and Kaniv Reservoirs. Identical Commissions were formed at the local level to deal with rivers in their locality. Thus Brovary and Vyshhorod Raions were concerned with the Desna River; while Voludarsky and Bohuslav were to monitor the River Ros. The situation at the Desna River was said to be so acute that the evacuation of the local population was prepared for.[28]

The initial task was to clean out the regular "structures" into which the thawed waters would pass. These structures consisted of some 5,500 drainage tubes under village roads, and gutters and sewers in the cities. Stepanenko made it clear that although the 30-kilometer zone around the damaged reactor was the first priority in preparing for spring floods, no precise prognosis on the level of the waters in that area had been ascertained. Possibly, he surmised, the spring floods will not be as strong as we imagine, "but events of the past year have taught us to pay attention to trivia; because of their neglect, we had to pay dearly."

Stepanenko, as a man on the scene, was very cautious. But in March 1987, the spokespersons for the central press and Moscow radio again adopted the official attitude that all would be well and that any fears about the effects of spring flooding were "groundless." Thus a representative of the Ukrainian administration of Hydrometeorology and Environmental Control, Petro Shendrik, having mentioned that the melted waters in the upper reaches of the Dnieper would carry 22 cubic meters of water into the Kiev Reservoir in 1987, compared to 17 cubic meters in 1986, was asked the following question:

> Let us return to the Dnieper. Residents along its banks are naturally worried about conditions around Chernobyl atomic energy station and in the Pripyat River basin. Won't the melted waters carry radiation from there into the Dnieper?

Shendrik's response was that there was no need for such fears. On the Prypyat and the other small rivers, networks of retentive hydrotechnical structures had been built to guard against the contamination of the Dnieper. The system, he stated, consisted of dams, settling tanks, underwater embankments and special dams on the Kiev Reservoir. But would *any* radionuclides break through this man-made protection system? The possibility, said Shendrik with remarkable confidence, had been eliminated and scientists had drawn up mathematical models to prove the fact. Even the snow itself was "practically clean." Thus, he assured listeners, "there is no basis for anxiety among the residents of the Upper Dnieper."[29]

A few days after this broadcast, Konstyantyn Sytnik, Vice-President of the Academy of Sciences of the Ukrainian SSR, and Director of the Botanical Institute of the Ukrainian Academy, informed readers of *Pravda* that "some increase of radionuclide content in the water is inevitable," thus contradicting Shendrik's statement about the mathematical model. However, Sytnik otherwise did not disagree with the earlier prognosis. He stated that in the Chernobyl area, about 6 cubic kilometers of water would flow into the Prypyat River. The river's banks had been reinforced, however, and he maintained that even if it burst its banks, the flood protection installations would prove adequate to hold back the flood. An icebreaker and demolition squads were standing by in case the river froze over.[30]

Looking at the flood preparation measures in more detail, it appears that the Soviet authorities were better prepared for this eventuality than for virtually every other aspect concerned with the aftermath of the Chernobyl disaster. Ostensibly, the reason behind such preparation was that spring flooding is something that occurs annually. Even without the nuclear accident, many of the measures applied would have been in force in the Chernobyl region. In the plant region, the bulk of the work was carried out by soldiers from the Engineering Corps of the Civil Defense of the USSR. They had the task of reinforcing structures already in place and building new ones in the floodlands of the rivers Prypyat, Uzh, Braginka and others. About 135 weirs and dams exceeding 40 kilometers in length had been prepared for the flood period. In each area, emergency supplies of sand, road metal and rubble were established.[31]

Even as late as May 1987, an "anti-flood watch" remained in place. Near the village of Lubyanka, in Poliske Raion, for example, the location of the Illya River, a large dam had been built. The "watch brigade" was employed around-the-clock for two 12-hour shifts. The brigade that had arrived from nearby Berdichev brought with them bulldozers, trucks, searchlights and various instruments that would be needed if the water were to break through the dam. The water was said to be rising and falling in unpredictable fashion as a result of a strange weather pattern of late snow and heavy rainshowers. A truck driver, interviewed in a newspaper account, declared that the reporters "should explain to the people in Moscow and Kiev that there is no radiation here in the rivers."[32] Does this imply that people believed that a significant quantity of radionuclides had penetrated the rivers?

Despite the above measures and the attempts to reassure the local population that the floods would not endanger lives, rumors abounded that the reality was otherwise. A Radio Kiev interviewer asked the Kiev doctor Tamara Lushchenko:

> Wasn't it [the radiation situation] worsened by the snow covering the area around Chernobyl and by the flooding? Certainly the snow water carried radioactive particles into rivers that had earlier fallen out into the soil and snow.[33]

Yurii Izrael, Chairman of the USSR State Committee of Hydrometeorology and Environmental Control, and a member of the 1986 Soviet delegation to the IAEA in Vienna, was asked to respond to the "wave of rumors" that the radiation levels rose suddenly in several locations after the melting of the snow. Izrael acknowledged that the increase had occurred, but offered the following explanation:

> If you speak of snow, then actually, where the radiation levels are located, through the melting of snow they may increase slightly. This is a measured effect, because a thick layer of snow—and this year you know there was a lot of snow—shielded the radiation that is found at the surface of the earth, and the snow was thus located between the radioactive contamination of the ground and the instrument....After the snow covering disappeared, we indeed measured several increases in radiation levels...but they were smaller than before winter.[34]

At the same time that Izrael offered his opinions (April 26, 1987), Vladimir Kolinko and Andrei Pralnikov of *Moscow News* were writing that finally "one can heave a sigh of relief." The danger from the spring floods had passed. Some radionuclides had evidently penetrated the waters, but not enough to prevent the Dnieper waters from being consumed by some 32 million people. The radioactivity of the Prypyat River and the Kiev Reservoir was declared to be at a "very low" level.[35] Nevertheless, although the main danger had evidently been averted, hazardous reconnaissance trips had to be made by helicopter to the 30-kilometer zone to test the waters.

For example, because of continuing concern over the Dnieper River, helicopters were taking off regularly from Kiev in the spring of 1987. The helicopters, manned by staff of the Ukrainian Center for Radiation Monitoring and the Study of Environmental Pollution and other hydrometeorological agencies conducting parallel tests, would land at the mouths of rivers on narrow strips of land. On one occasion, which appears to have been typical, the heads of the Ukrainian Center's two laboratories jumped out of the cabin, ran down the hillside to the water's edge and filled polythene containers with water samples within a few seconds. They then rushed back to the cabin. Once a large wooden box had been filled with samples, then the helicopter returned to Kiev.[36]

The article provides an example of the dangers and limitations involved in measuring the state of the water in the 30-kilometer zone. Whether the main danger had passed by the spring of 1987 is a moot point. As will be seen below, the vast majority of villages in the zone were still considered unfit for habitation, and one reason for the delay in repopulation was uncertainty about the state of the water. There are no independent assessments of the state of the water that was subject to radiation fallout, in contrast to the instruments and equipment of the nuclear plant itself, which have been open to limited foreign inspections. What can be said is that the Soviets were prepared for spring flooding, but that some radionuclides entered the water supply. The Dnieper was unfit for drinking in the summer of 1986, but according to Soviet reports, the main danger ended after the spring floods of 1987.

The Food Chain

Before examining in detail the impact of Chernobyl on agricultural production in the 30-kilometer zone, it is pertinent to conduct a brief survey of current knowledge about the impact on the food chain. Various Soviet accounts have praised the precautionary measures adopted by the authorities in the first days after the event. Anatolii Romanenko, for example, commenting that the actual doses of radiation received by "tens of thousands" of people were 2-3 times less than originally forecasted, attributed the lower levels to the effectiveness of the protective measures in the "first days."[37] As will be shown, such measures were adopted, but somewhat belatedly in most areas.

Similarly, Aleksandr Kondrusev, the head of the Main Sanitary and Epidemiological Department of the Ministry of Health of the USSR, interviewed by TASS on the first anniversary of Chernobyl, stated categorically that "I have every reason to say that the population did not receive any contaminated food." He maintained that a rigorous radiation control system had been established to ensure that the food and water in the Belorussian and Ukrainian republics met sanitary and hygiene requirements.[38] The reality, however, was quite different.

For example, in July 1986, a report from the Belorussian side of the 30-kilometer zone stated that the advice of medical workers about food intake was not being fully implemented. On the contrary, the local authorities had manifested "inefficiency and inertia" in carrying out the needed protective measures.[39] The weekly magazine published by *Izvestiya*, called *Nedelya*, referred in an issue of September 1986 to the lack of information in Ukraine about which food products were prohibited and which could be eaten by residents in areas of increased radiation background.[40]

Even in the spring of 1987, the precautions varied widely from area to area. Thus an account in a Ukrainian newspaper at the Chernobyl nuclear power plant itself observed that people were again fishing in the canal that surrounded the four reactor units and that "such activity has not been observed for a while."[41] Yet on the Belorussian side of the zone—and at a considerably greater distance from the damaged reactor—fishing was still forbidden. Thus in mid-April 1987, I.N. Nikit-chenko, the Deputy Chairman of the Belorussian State Agroprom, a Doctor of Agricultural Studies, warned that people should not fish in the reservoirs or pick mushrooms in the southern raions of Gomel Oblast "this year" (i.e., for the duration of 1987).[42] Such discrepancies in precautions from area to area were a familiar feature in the early days after the accident. That there were such differences a year later is astonishing. It reveals the uncertainty over what steps to take against radiation fallout at the local level.

After the accident, the Ministry of Health Protection of the Ukrainian SSR began to apply measures against radioactive iodine-131. A considerable quantity of milk was destroyed. Some milk was considered unsafe for consumption, but was preserved for the preparation of butter and cheese. The controls on milk consumption lasted approximately from May 1 to August 1. It is by no means clear that they were officially rescinded on the latter date, but by this time the authorities were declaring that milk and dairy products could now be safely consumed.[43] During the three-month period, "clean" food products had to be brought into some villages from the outside, particularly the three Ukrainian villages within the zone that had not been evacuated, and in areas of Belorussia in which pockets of high radioactivity were discovered well outside the 30-kilometer limits.

One example of the latter was Khoiniki Raion in Belorussia. Here, it was reported, routes for mobile stores were laid out for every village and hamlet. These stores brought bread and goods needed for daily use (but evidently not milk). Throughout the Belorussian SSR, brigades of sellers and cooks were formed to work in the polluted raions by the shift method. The report about this activity in *Izvestiya* revealed that a significant number of those working by this method—which involved being separated from wives and families—did not volunteer but were rather assigned to their tasks.[44] In many areas, the local villagers did not receive adequate supplies of milk and meat, and lacked basic storage facilities—such as refrigerators—in which to store such items.[45] So how did these people supplement their diet? The answer is that they either went hungry, or that they consumed contaminated products. And there is no doubt that both occurred.

Thus Ian Waddington of the World Health Organization pointed out during an interview early in 1987 that Soviet officials had not succeeded in controlling consumption of meat, milk and vegetables in the rural regions of the zone around the defunct reactor: "They [the Soviets] never did manage to control the milk and local consumption of meat and vegetables in the inner exclusion area of 130,000 people." Even beyond the zone, Waddington maintained, many must have taken in considerable levels of radiation through the ingestion of food.[46] One can deduce that the most dangerous period for such consumption was between April 26 and around May 21, 1986, when the final period of evacuation was completed.

Some products would also have found their way onto local markets. Other than official reports, which invariably deny that any contaminated products were sold, there is no means of measuring the extent of such sales. As one Soviet source noted, however, "a situation was possible in which a dishonest person decides to carry out a sale."[47] On December 14, the correspondent for the *Washington Post*, Gary Lee, observed a vendor at Kiev's central market shouting "Apples, apples without radiation," whereupon she was asked by shoppers how she could prove such a claim. Despite three dosimetric checks for each item of food, Lee noted that there were still fears that some contaminated food would reach the market, and the situation may have been worse in smaller settlements.

By the late Fall of 1986, a ban placed on the sale of blackcurrants in the city of Kiev had been lifted. By the summer of 1987, products such as honey, bilberries, currants, cucumbers and peas were all available at Kiev markets. Leonid Ilyin stated that the Ministry of Health Protection of the USSR had confirmed that in "recent months" (early 1987), there had been no reason to place restrictions on the sale of any products.[48] The comment is confusing, however, because it does not mean that there were no such restrictions, only that practice had proved that they were not necessary. In April 1987, two Kiev scientists were asked about the problems caused to nutrition by the accident. They replied that the fruits of blackcurrants, redcurrants, plums, and cherries could be consumed without hesitation. But within the 30-kilometer zone, there were still restrictions on the grazing of cattle and the consumption of milk.[49]

Ilyin also informed a Congress of International Physicians for the Prevention of Nuclear War that in the areas north of the Chernobyl plant, 270,000 samples of foodstuffs had been analyzed, of which almost 30,000 turned out to be contaminated,[50] an unusually frank admission from the Soviet scientist noted for his conservatism in reporting on the effects of the disaster.

Sheep breeding in the Chernobyl zone, which was carried out in only a few areas, was banned after the accident. Cattle-breeding continued, however. After the accident, the cattle were evacuated along with the collective and state farms within the zone. The radioactive fallout was widely dispersed on the cattle-grazing lands and forced many dairy farms to change their function to year-round indoor cattle maintenance. The cattle were then placed in pens and provided with "generous supplements" of salt and minerals. The cattle were fattened for milk in two stages. In the first six months, they continued to eat the often-contaminated local feed products, after which they were transferred to "clean" grazing areas where the grass was at least 15 centimeters high. Reportedly a further six months proved adequate for the animals' organisms to rid themselves completely of radionuclides. In the meantime, the contaminated cattle were not allowed to graze in forests, where there was a further danger of cesium contamination and their milk had to be sent for processing.[51]

How is one to summarize these findings and Soviet measures in food protection generally?

In brief, the chief short-term effects on the food chain occurred in the first hours and days after the accident. By the time Ukrainian Health Minister Romanenko gave the first health warning to the Ukrainian population on May 5, 1986, the main damage had been done. Thereafter, the measures applied appear to have been exemplary in cities such as Kiev, with its 23 markets, but were less satisfactory and often varied considerably from area to area elsewhere.

On July 17, 1987, TASS announced that Kiev cooperative markets could now sell cottage cheese, sour cream and butter that had been produced on the personal plots of collective farm households. All dairy products were nonetheless subject to tests for radioactivity. This signified an end to major post-accident restrictions. Henceforth, it seems that the only precautionary measures for food consumption applied to the 30-kilometer zone.

Impact on European Agriculture

Before examining Ukrainian agriculture after Chernobyl, it is useful to look briefly at the European scene by way of comparison. Much of our early knowledge about the accident's impact came from reports from Scandinavia and regions in Western Europe that felt the effects of the radioactive cloud. We will look at two aspects that have been fairly well documented: the effect on the reindeer herds of Scandinavia; and the influence on sheep breeding in mountainous regions of the British Isles. In both cases, the accident has had longlasting and serious effects.

The Swedish government has stipulated a maximum limit of 300 becquerels per kilogram for the concentration of radioactive products in reindeer meat. In Norway, the maximum level is 600 becquerels. However, the fallout of cesium-134 and cesium-137 from the Chernobyl accident brought levels far in excess of the official limits. The reindeer, which form the mainstay of the existence of the nomadic Lapp tribes of Sweden and Norway, feed on lichen, which is extremely sensitive to radioactive fallout because it takes its nourishment from the atmosphere. Following the first post-Chernobyl rainfall, the lichen soaked up cesium, which was then eaten by the herds of reindeer.[52]

The NRC's report on the Chernobyl accident states that the level of contamination in reindeer meat in Scandinavia reached 20,000 becquerels per kilogram, and that concentrations well above maximum limits will persist for several years.[53] A Western news agency reported that whereas the average radiation level of cesium-137 in the worst-affected areas of Sweden averaged between 1,000 and 15,000 becquerels per kilogram, in Norway, levels were up to 40,000 becquerels per kilogram, or 66 times higher than the official limit. In Sweden, high radiation fallout affected about 90,000 of the 270,000 head of reindeer (or the vast majority of those about to be slaughtered), drastically curtailing the lifestyle of the 15,000 Lapp population of the country; moreover, 80% of Norway's 80,000 head of reindeer were said to be seriously affected by the fallout of about 1 kilogram of radioactive cesium (about 6% of the total emission) after the Chernobyl accident.[54]

The consequence was the burial of 75% of the reindeer slaughtered in Sweden or the sale of the reindeer meat as animal feed.

According to one source, the Lapps themselves soon began to show the effects of eating reindeer meat. Thus whereas the average resident of Stockholm manifested a cesium level of less than 50 becquerels per kilogram in the summer of 1987, an examination of some 200 Lapps revealed that their levels had risen from 10,000 to 30,000 becquerels between March and June 1987, as reported by Goran Wickman of Umea University's Radiophysical Institution. He noted further that reindeer meat would probably not attain acceptable concentration levels for "several decades."[55] Chernobyl has imperiled the lifestyle of the Lapp race and may possibly have destroyed the entire reindeer industry that has long predominated among the Lapp herds, where even the vocabulary of the Lapp language is made up of numerous words derived from reindeer. Thus Jorgen Bohlin, the legal advisor with the Lappish National Union, was reported as saying that the reindeer-herding Lapps in Northern Sweden's most affected areas might consider taking up offers to resettle in areas such as British Columbia, Canada.[56]

The radioactive cesium that affected the reindeer herd of Scandinavia also had its impact as far west as the British Isles. Early in May 1986, there were heavy thunderstorms over the mountain regions of Cumbria, North Wales and western Scotland that resulted in the contamination of between 2 and 4 million sheep. Even with the passage of time, the levels of cesium remained higher than anticipated, probably because the cesium element moved from the grass into the soil more quickly than had been foreseen. According to one source, by February 1987, when the contamination of the Cumbrian regions had fallen to 10-20% of its post-accident level, 150 farms were still registering high cesium levels in spring lambs. In North Wales, 315 farms were said to be similarly affected.[57]

In August 1987, new restrictions were imposed on 69 highland farms in Scotland, involving 124,000 head of sheep. Altogether at this time, the embargos on the slaughter of sheep for consumption in Britain encompassed 564 farms, with 560,000 sheep. The report, in the *Financial Times*, cited "experts" from the British Ministry of Agriculture as stating that the farms in Scotland possessed poor peat soils that did not have the mineral content with which to ward off radioactive cesium.[58] These two examples from distant parts of Europe demonstrate the pervasiveness of the cesium element from the Chernobyl reactor. It has lasted much longer than had been anticipated. Moreover, since the majority of its fallout occurred in the region around the plant itself, it is evident that the effects of the fallout on Ukrainian and Belorussian agriculture were even more severe than in Scandinavia and Britain.

Agriculture in the Special Zone

It has been stated above that the northern part of Kiev Oblast was not a significant area for agricultural production. As readers of Soviet newspapers often pointed out, it could have been abandoned permanently without any major effect on the economy. As Valerii Legasov noted at the World Energy Conference in Cannes, France in October 1986, the country has lost larger and more significant stretches of farmland from the construction of hydroelectric dams than from the ground made useless as a result of Chernobyl.[59]

The report released by the United States' Central and Defense Intelligence Agencies in March 1987 also noted that the accident's impact on agriculture was minimal. Of the 1,000 square kilometers affected, more than 50% were made up of forests and swampland, constituting "a miniscule share of total Ukrainian farm output."[60] Why then should lives have been put at risk in the hazardous tasks of decontamination? Why could not the land have been left fallow, forming a dead zone around the nuclear plant, an eternal reminder of the damage wrought by the uncontrolled atom?

The official response to such comments has always been that every effort should be made to bring the land back into agricultural usage. Why? The main reason cited is that it was essential to eradicate as far as possible the damage caused by man to his natural environment:

> By allowing the accident to happen at the nuclear power plant and land to become contaminated, we became indebted to nature and to our descendants. That is why we must do everything possible to redress the consequences of the accident. The opportunities for such a course are considerable. We have the knowledge, the means of application, the will of the people. In addition, there are thousands of evacuees who want to return to their native homes.[61]

Chernobyl forms a chapter in Soviet history that in official writing will be recalled as a success, even though serious criticisms have emerged in the period of glasnost. Its effects must therefore be eliminated. In addition, the nuclear plant itself has not been abandoned, even though its employees can no longer be housed permanently inside the zone.

An uninhabited desert around the plant appeared to the authorities to be an alternative that was simply inconceivable.

The losses to the harvest were relatively small. In an article of October 1986, the Chairman of the "Gorky" collective farm, Makariv Raion, Kiev Oblast, estimated that as a result of the accident, the Kiev Oblast would be "deficient" by some 25,000 tons of grain, 70,000 tons of potatoes, 30,000 tons of milk and 1,000 tons of flax.[62] In terms of republican output, only the losses to the potato crop are measurable and even then would have constituted about 0.35% of total production for the 1986 year. However, the crops would have provided for about 90,000 people who were eventually evacuated; while the future area, it is estimated, will eventually have to provide for a somewhat smaller population.

In the summer of 1986, the Soviet authorities acknowledged that the wheat in the fields around the damaged reactor was contaminated and could not be harvested. Nonetheless, it was felt that means of decontaminating this wheat should be explored. An agronomist from the Botanical Institute of the Ukrainian Academy of Sciences, D. Grodzinsky, suggested several ways to remove from the soil radioactive cesium and strontium elements which would otherwise move through the roots of plants into animals and humans.[63]

One notion was to cultivate plants such as lupins that can absorb radioactive elements from the earth, and then dispose of them as nuclear waste. But the method is time-consuming and costly. A second option was to irrigate the land with a mixture of chemicals such as soluble calcium and water, which would seep into the soil, draining radioactive elements to a level safely below the roots of the plants. A third option, in Grodzinsky's opinion, was to fertilize the land with lime and ashes, and then turn the topsoil, after which the radioactive elements would not rise to the surface again. However, he maintained that close to the reactor, it would be wisest to leave the land fallow for a time. Farmers might consider changing the nature of their crop production from wheat, which is highly sensitive to radioactive substances, to rye and potatoes which absorb less of the radioactive fallout and can be safely processed.

By the spring of 1987, the change of direction in agricultural production had been confirmed, but the prognosis was much more optimistic. This became evident from an interview in the Ukrainian press with N. Arkhipov, a member of the State Committe for the Utilization of Atomic Energy, and head of the Laboratory for Land Reclamation in the Ukrainian SSR. Arkhipov had led a group that came to Chernobyl in May 1986 with the State Agroprom (Agroindustrial Committee) and had conducted experiments on the contaminated areas to ascertain whether they could be used in the future. At that time, their conclusions were that the zone had suffered a severe setback. "Today," however, "we are sure that the greater part of this territory can be made arable again."[64]

Following the treatment of farm lands with chemicals, Arkhipov continued, the contamination of the biological environment had fallen by 20-30 times, which led to the brighter outlook for the future. He also maintained that the radionuclides discharged were relatively immobile because they are insoluble in water and thus cannot spread rapidly through the chain soil-plants-animals-man. Instead, they were more dangerous where they had landed on the ground and could be carried by the wind, together with dust. The answer to this problem, in his view, was simply to plow them into the soil, where they would be held and decay over the passage of time. Although the method seemed dubious—in effect it will contaminate the soil for three decades—it had evidently been tried out with some success in the Polissya region, where the harvest turned out to be 100 times cleaner than in neighboring zones.

But would this method prevent the movement of cesium and strontium, which, as noted in the European areas, proved to be highly mobile? Arkhipov felt that in these cases some agent was required to bind the isotopes into the soil. Thus in areas contaminated by cesium and strontium, the soil had been treated with lime and zeolites. The latter are special types of clay that increase the absorption properties of the soil. According to another source, they block and imprison the molecules of strontium and cesium and hinder their advance toward the plant roots. It was necessary to deliver the zeolites from Ukraine and especially from Georgia.[65]

It was important, Arkhipov stated, to raise constantly the yield of crops, because with each increase, the radionuclide content of these crops decreased. He then made the most surprising comment of the interview, which was that despite all previous calculations to the contrary (especially by scientists in the West), the radiation levels would permit the cultivation of at least half the territory of the 30-kilometer zone by the spring of 1987. The main obstacles would stem not from radiation, but rather from social, economic and psychological factors, i.e., whether people would be willing to return home and whether they would have the facilities on the spot to enable them to do so.

Turning in more detail to the "protective" measures applied to the soil in the 30-kilometer zone, it becomes evident that Arkhipov's account was necessarily simplistic, geared to calming the anxieties of his reading public. More detailed accounts appeared elsewhere, and Soviet scientists appeared prepared to divulge information that they were not willing to provide for other aspects of the post-accident situation, such as the level of radioactive fallout into the water supply. We will provide one example from Ukraine and one from Belorussia.

According to two prominent Ukrainian scientists, the first task was to neutralize acidic soils, such as turf-podzol, forest gray and podzol-type chernozem, using slaked lime and limestone. Excellent results, they stated, were obtained by using dolomite flour, which when introduced into the soil eliminates deficiencies of magnesium at the same time as neutralizing the soil. The introduction of lime materials was carried out twice, before and after the main deep digging of the soil or plowing. The soil had to be "chalked" when the hydrolytic acidity was twice the norm. The approximate doses of lime added were from 600 to 1,000 grams per square meter in the Polissyan zone with its acidic podzol soils, and from 100 to 200 grams on chernozem soil.[66]

The scientists emphasized the importance of preserving the accepted ratios in the soil of its basic plant nutrients: nitrogen, phosphorus and potassium. A surplus of nitrogen in the soil, they noted, would lead to an increased accumulation of radionuclides by plants, and hence phosphorus and potassium fertilizers should be introduced in doses sufficient to create a slight surplus relative to nitrogen. For every unit of nitrogenous fertilizer substances, one should have 1.5 units of phosphorus and 1.5-2.0 units of potassium.

A similar complication arose with watering. Although irrigation and watering raise the productivity of plants leading to a "carbohydrate dilution," and the decrease in the concentration of various toxins therein by 1.5 to 1.8 times, an excess of watering can be harmful. The scientists pointed out that "supplementary moistening" raises considerably the availability of radionuclides for adoption by the root systems of plants. The goal was to introduce the wide application of soil-protective agricultural systems.

The situation in Belorussia was subjected to a thorough analysis by I.N. Nikitchenko, Deputy Chairman of the State Agroprom of the Belorussian SSR, a doctor of agricultural studies. Having outlined the methods described above by the two Ukrainian scientists, he provided more information on the changes in crop production as a result of the accident. First, he noted that the amount of radionuclides transferred into plants depends on the mechanical state of the soil, its physical and chemical properties, and on the varieties of crops. Thus radionuclides were absorbed more in lighter and acidic soils and less in heavier and neutral soils. Also, plants that have accumulated calcium—specifically vegetables—take in more strontium, whereas in crops such as buckwheat, which reveal a preference for potassium, the cesium intake is higher. Further, late ripening varieties of crops would have taken in 1.5 to 2.0 times fewer radioisotopes than the early ripening crops.[67]

Nikitchenko declared that depending upon the contamination, the composition of the soil, the cultivation of the fields before sowing and other indicators, the authorities recommended the cultivation of winter and spring varieties of wheat, oats, barley, winter rye, flax, sugar beets, cucumbers, tomatoes and cabbage. Buckwheat and corn were to be cultivated only on sandy and clay soils that had low levels of contamination. Having described measures for collective and state farmers, Nikitchenko then provided some directions for those working on personal household plots and "amateur gardeners." These were reportedly embodied in specially drawn up instruction leaflets that had been distributed among the Belorussian population within the 30-kilometer zone (this author has not come across any evidence of such leaflets on the Ukrainian side of the zone, but this is not to say that they did not exist).

The leaflets stated, for example, that when the earth gets warm, one must dig it up with a shovel as deeply as possible, turning the clods over without breaking them up, and then level the surface and pack it. The garbage collected from the previous year, such as old leaves, fallen branches and dried weeds, should be collected and buried (not burned!) at a depth of not less than one meter, in some corner where it was not planned to plant trees or berries. The top of the burial pit should then be covered by 20-25 centimeters of clean soil. If this had not been carried out in the Fall of 1986, then the soil had to be "chalked" with superphosphate and potassium chloride (or sulfate), and zeolite could also be applied. In each individual plot, 50-60 kilograms of lime were to be added. It was emphasized that wood and peat ash must not be used as fertilizer.

On the garden beds, according to the directions, the farmers should grow root crops. Peas, haricots and beans were not recommended. Tomatoes could be grown if planted in holes filled with peat that had been taken from under the surface of the soil. When growing cucumbers and the vegetables had appeared above the surface of the soil, it was advisable to elevate them so that the bottom part of the cucumber was no longer in contact with the soil. Even better, it was stressed, they should be grown "under a film" in clean soil if possible. As noted above, it was considered inadvisable to pick mushrooms or to fish in the local reservoirs in the southern raions of Gomel Oblast. If the local residents required further elucidation on any of the above points, they were advised to talk with farm specialists or visit the local village Soviets.

One should differentiate between advice and orders in the above leaflets. Without actually perusing such a leaflet, it is impossible to know whether a distinction was made in the Belorussian zone between what one *should* do and what one *must* do. In any event, most of the directions were probably unenforceable and depended on the radiation fallout from area to area. However, it is noticeable that in the Ukrainian press, those scientists and government spokespersons interviewed were notably less cautious than those in the Belorussian press. This is paradoxical since the Ukrainian territory made up the majority of the 30-kilometer zone. Although the information about soil treatment was similar in both regions, the advice given to local farmers appeared to be quite different.

Take, for example, an article in the Ukrainian newspaper *Robitnycha hazeta*, dated May 20, 1987, in which a Kiev agronomist, A. Kalenich, was being interviewed. In the first place, he concentrated almost exclusively on the republic as a whole rather than the 30-kilometer zone: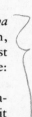

> ...on collective gardens in all the territory of the republic, the radiation situation permits the cultivation of potatoes, vegetables and fruit crops without any restrictions. The content of radionuclides in them will essentially be below the permissible norms. No special measures are needed for market-garden plots.

Kalenich was informed that the newspaper had received letters asking about individual protection for work on garden plots. In the "overwhelming majority" of oblasts of the republic, he stated, this is absolutely unnecessary. Only for the gardens of the raions in the 30-kilometer zone, "*it is not a bad idea*" to wear respirators or bandages made up of several layers of gauze. Such understatements, combined with the emphasis on the republican rather than a zonal situation tended, despite indications to the contrary, to minimize the danger in Ukraine's 30-kilometer zone.

A month later, Leonid Ilyin was asked what he would recommend regarding agriculture on farms and cooperative gardens, "of which there are so many around Kiev?" The preposition "around Kiev" could in theory have been applied to the 30-kilometer zone, but Ilyin's response applied only to Kiev itself and with its nonchalant dismissal of the possibility of any danger in the capital, seemed to infer—because of the nature of the question—that there were no dangers in the Kiev area as a whole:

> Let Kievans go out in the fresh air, swim in the Dnipro, not restrict their intake of greens, milk or other products. Let them work on the collective farms—this can only be beneficial to their health. Whoever has doubts, let him measure the radiation contamination of the products himself; there are dosimetry points located in the area.[68]

Ilyin's comments were a virtual paraphrase of Romanenko's original health warning of May 5, 1986, with the difference that the negative had now been changed to the positive case. In fairness, however, it should be borne in mind that Ilyin was concerned also about radiophobia in certain areas and wanted to encourage people to eat properly. Nonetheless, the contrast with the warnings given out in Belorussia is striking.

If warnings to the population in the Ukrainian regions of the 30-kilometer zone, and polluted areas beyond that zone were not carried in the press, did restrictions in fact exist in this zone? The answer is yes. Of the 25 administrative regions in the Ukrainian SSR, 3 had been singled out for careful attention, namely the Kiev, Chernihiv and Zhytomyr Oblasts. According to one source, the USSR Agroprom recommended in the summer of 1987 that the Chernobyl region itself be given the status of a state reserve, but no formal decision had been taken "even though it is high time." Nevertheless, agricultural production was under close state supervision (by which one assumes not only the State Agroproms, but also military supervision) in the Chernobyl and Poliske Raions of Kiev Oblast, and parts of the Ovruch and Narodych Raions of Zhytomyr Oblast, which were well outside the 30-kilometer zone. However, although the state had made recommendations as to which crops should be cultivated, "they are not compulsory."[69] This implies that there were limitations on how far local agriculture could be controlled. It was impossible, for example, to monitor every private plot on every collective farm household.

One means of monitoring the effect of radiation on plants and fruits was the "Kompleks" radiobiological laboratory in Prypyat (see Chapter 6), which was a normal hothouse before the accident occurred. After the explosions, however, the roof of the hothouse was smashed and radioactive particles entered the hothouse, Subsequently, it was transformed into a scientific experimental center to examine the effects of relatively small doses of radiation on fruit crops. The Kompleks was cited as proof of the "safe" levels of radiation in fruit, and also as evidence of the revival, or continuing function, of the city of Prypyat, which remains for the most part uninhabited. But it has also led to optimistic statements about the level of penetration of radiation in plants, which in turn has influenced the measures adopted. Thus Aleksandr Yakovlev, the Chief Agronomist of the laboratory, stated that it has now been established that although the root system of the plant may accumulate radionuclides, the fruit still remains "clean."[70] The implication, in short, is that the soil can be contaminated without adversely affecting the fruit produced.

Having focused on the methods of radiation prevention in the soil, the attention to the dust on the soil surface should also be mentioned.[71] This problem was also studied at the Prypyat hothouse, but the chief measure that was eventually applied to combat the longlasting problem of dust—rather than the initial measures taken after the accident—was elaborated at the Kolsk section of the USSR Academy of Sciences. The method was as follows. The main difficulty stemmed from sandy soil that was subject to wind erosion. This soil was covered with a thin layer of water, to which plant seeds were added by basic hydro-seeding machines. The seeds were contained at the surface in a film. As they grew, they formed a sod-containing stratum, thereby preventing the wind erosion of the soil.[72]

The above precautions and preventive measures are cited as examples of the effectiveness of the anti-radiation measures applied in the wake of the Chernobyl disaster. Yet how were they implemented in practice? And did they prevent the spread of radionuclides into the soil? Have they enabled the revival of agriculture in the zone?

The evidence suggests that many of the problems encountered appeared almost insurmountable. In working to combat dust, for example, much depended on the use of tractors with hermetically sealed cabins. In the Belorussian zone, however, as late as mid-April 1987, beyond the time when a Ukrainian scientist was declaring that more than 50% of the entire zone was being cultivated, not one machine had arrived from the tractor factory entrusted with producing the machines in Minsk. None would be available until the second half of 1987. The necessary combine machines had also not arrived from Gomel.[73]

As for the field cultivation, before haymowing and pastures could be restored, it was considered essential to replenish the seed surpluses of the cereal grains. In Mogilev Raion, Gomel Oblast, an additional surplus of over 500 tons was needed, but by April 1987, only 72 tons had been assigned, and not all of this amount had been delivered to the raions. The delivery of zeolites was "unfortunately delayed," and in 1987 the farmsteads had received "scarcely any." Why had these delivery bottlenecks occurred? The official response was that it was a consequence of a lack of coordination, irresponsibility, the habit of "talking around questions" rather than simply carrying out one's instructions. It would be a good idea, noted Nikitchenko, to keep in mind that all these matters "concern raions that we call the zone of special attention!"

The levels of contamination of the soil and plants have also been reported as somewhat more acute than suggested by some of the interviews cited above. Thus Oleg Pavlovsky of the Institute of Biophysics of the USSR Ministry of Health Protection stated that one year after the accident there were "unfortunately, still problems." In some types of soil, he revealed, radioactive substances were entering plants "especially actively." Scientists were trying to combat this development with fertilizers and special chemical reagents, but he did not state whether such methods were having any effect.[74]

In many areas, even in the spring and summer of 1987, the initial radioactive fallout had left a surprising number of areas contaminated, preventing the possibility of the return of the original inhabitants and necessitating exceptional attention. Colonel-General Vladimir Pikalov commented that the soil around the reactor was still contaminated. Several "pockets" in Belorussia were showing "higher than admissible" radiation readings, he noted, and the soil was polluted with cesium, strontium and plutonium radionuclides that were being flushed out "at a very slow rate."[75] A spokesperson for the Kiev region said that there were various "dirty spots" on the Ukrainian side of the zone, during an interview of mid-June 1987.[76]

Those attempting to carry out farming in the zones around the Chernobyl plant were facing serious dilemmas. For one thing, the area was contaminated and yet they had to provide for themselves, in addition to following the various rules for grazing their cattle and controlling their produce. In one raion that appears to have been typical, the local villages relied on condensed milk and the occasional delivery of poultry from "clean" zones. However, they almost never received any meat products, mainly because the roads were too bad to ensure delivery by anything other than a horse and cart. Those in distant areas received milk only once per week once the winter was over, and this milk was invariably sour by the time it arrived. There was not one refrigerator in the entire raion. For milk production, there was only one isothermic machine, but at least five were needed. However, the cattle from private farms had to be grazed not on the safe cultivated pastures, but in the forests and hayfields where radiation contamination was up to five times higher. Although grazing on cultivated land existed in theory, "from paper to action is a great distance."[77]

The above example illustrates well the problems in agriculture within the danger zone, in addition to the difference between theory and practice in protecting against radiation fallout. The enormity of the accident has effectively nullified any possibility of conducting normal agricultural work in the Chernobyl zone. Those farmers who have been assigned such a task are working at grave risk to their health, not only from external contamination, but also from contaminated crops and the lack of basic foodstuffs (and clean water, too, although that is a moot point).

Again, the Ukrainian authorities have been anxious to play down the problems involved, at least in their communications with the Ukrainian public. While there may have been some valid reasons for adopting this attitude—to prevent panic, for example—the overall impression one gains is that the local villagers were never fully cognizant of all the dangers involved. This will become more evident (in Chapter 4) from our examination of the lives of some of those evacuated from the reactor area. Having examined the relative insignificance of Chernobyl Raion as an agricultural source, and the lengthy duration of the strontium and cesium radionuclides, the decision to try to return the area to agricultural usage must be questioned.

However, the current Director of the Chernobyl nuclear power plant, Mikhail Umanets, stated in June 1987 that the "chief task" of the workers in the State Agroprom's section in Prypyat was to "recultivate this land and restore its place in the national economy." By July, TASS was announcing that over 75% of the land in the 30-kilometer zone was now under cultivation, i.e., 212,000 out of 282,000 hectares. Decontamination work was being carried out only on "low-yield fields" with sand and clay loam soils.[78] Since in July 1987, about 66% of the evacuated villages were still uninhabited, most of the agricultural work must have been carried out by the military and civilian defense forces. This does not lessen the impact of such a momentous decision. The dead zone had been revived, against all odds, and seemingly against all reason.

3 The Economic and Political Repercussions

This chapter will focus on the economic and political impact of the accident on the Soviet Union, and on Soviet Ukraine in particular. Its conclusions are tentative because these repercussions are still continuing, and it will be some time before a definitive assessment can be made of the accident's impact on Soviet nuclear energy policy. We will limit ourselves to events that have occurred since the summer of 1986. The Soviet nuclear energy program and its Ukrainian component have been described in detail elsewhere. Here, therefore, it will suffice to look at those programs as they developed after and in the light of the accident. Similarly, the progress but not the background of the East European nuclear energy program will also be noted.

The Electricity Crisis of 1986-7

At the outset, the Chernobyl accident did not appear to have had a major economic impact on the USSR for the following reasons:
1. As noted above, the accident occurred in one of the least productive agricultural regions of the Ukrainian SSR that consisted primarily of forests and swampland. The loss of these lands, even for a short period of 1-4 years would not have a significant impact on Soviet

agricultural production unless the Dnieper water supply system had been seriously contaminated, which was evidently not the case.

2. The vast transfer of personnel that was carried out at the state's expense actually made some economic sense. In the southern areas of Gomel Oblast of Belorussia in particular, the movement of people to the areas of acute labor shortage in the northern part of the oblast helped to rectify a longstanding economic problem. Some evacuees ended up in the Donbass region of eastern Ukraine, which is also an area of labor shortage. Some personnel at the nuclear plant were also transferred to new Ukrainian plants under construction, such as that at Khmelnytsky in western Ukraine.

3. The accident led to the destruction of one RBMK-1000 reactor and the shutdown of three others at the Chernobyl plant for periods of 1-2 years. But the accident occurred in late April when the electricity load in the Soviet Union and Eastern Europe was approaching its period of least usage. Moreover, thermal power stations worked through their summer maintenance periods to compensate for the sudden loss of electric power production at Chernobyl.

4. The 1986 year was to have seen the operation of four new reactors of the water-pressurized VVER type, all of 1,000 megawatts' capacity. Three of these reactors were to be in Soviet Ukraine: Rovno-3, Zaporizhzhya-3, and Khmelnytsky-1. The other was Kalinin-2. In addition, new thermal power plant units were planned at Kiev and Gomel stations, both within the area affected most by the loss of the Chernobyl plant.

5. It should be recalled that although the size of the Chernobyl plant rendered it a significant component of Ukrainian electricity production, on the all-Union level, it was less important. It is estimated that at full power it accounted for only about 1.3% of total electricity production in the USSR, and it should be emphasized that a portion of the electricity produced there was being transferred to the East European grid (to Hungary in particular) by means of a 750-kilovolt line that began in Vinnytsya Oblast of Ukraine.

For all the above reasons, the economic impact of Chernobyl did not appear to be major. At most, it was expected to cause minor short-term problems in electricity production that would affect certain industries which had depended upon the new 1986 reactors to meet the increased demands for electricity in the Fall of 1986. Yet the reality proved otherwise. A serious crisis situation emerged in the late summer and Fall of 1986 and lasted until the spring of 1987. Why did this occur?

In the first place, the accident brought into question the viability of the RBMK-type reactor. The short-term effect was that retrofittings were made to the 14 Soviet reactors of this type, which necessitated their temporary shutdown. It is not clear when the retrofittings began; possibly they did not begin immediately after the accident because it was not clear initially that the reactor's design was partly responsible for what had occurred. It does seem from the evidence available, however, that all the RBMKs were shut down simultaneously, thereby depriving the Soviet Union of not 8-9%, but around 55% of its nuclear-generated electricity as the summer of 1986 drew to a close.

Thus in late September 1986, Ye. Petrayev, the head of the integrated grid system in the USSR, was interviewed by a Soviet newspaper. The ostensible reason for the interview was a letter to the newspaper from a Moscow engineer, V. Dubovik. The latter, evidently a regular correspondent on power affairs, noted that since the middle of 1986, the frequency in the country's power system had been falling. In Dubovik's view, the Chernobyl accident could not have been responsible for the sudden drop in power because for a month after the accident, the frequency had remained normal, and "then for some reason dropped sharply."[1]

Petrayev responded that the power situation was indeed tense, but that there were many causes for the predicament. The loss of the Chernobyl reactors had been offset by mobilizing spare capacities at other stations:

> But let us recall the Government Commission's findings on the causes of the Chernobyl accident. In addition to staff negligence and violations of station equipment operating procedure, design failures were also noted. Thus this shortcoming is inherent in some other atomic energy stations fitted with similar equipment. Surely even the slightest risk of another Chernobyl cannot be tolerated? For that reason a whole package of preventive repair measures was undertaken at several atomic energy stations. To that end, a number of power units were taken out of operation and are now undergoing work to increase their safety and reliability. The loss of these capacities, albeit temporary, is not easy to make up.

Petrayev did say, however, that the most important factor was the quality of the repair work. It would be wrong, he declared, to deem a unit repaired, have it work for one month and then break down again.

But was he not alarmed that the number of stations undergoing modifications was much larger than planned and that winter was quickly approaching? "One cannot fail to be alarmed!" replied Petrayev. The situation was worse than anticipated.

Around the time that this interview appeared in the press, other reports were also indicating an acute situation in the power supply of the country. On September 29, 1986, *Pravda* announced that the Soviet Union was facing a shortfall of energy for the winter because of the Chernobyl disaster. But other factors were also playing a part, such as the water shortage—the low levels in rivers were adversely affecting the performances of hydroelectric stations, which account for about 17% of all electricity production in the country—and delays in work on constructing "three" nuclear power plants. The newspaper was referring to the three new Ukrainian reactors.

In early October, an editorial appeared in *Izvestiya* about the predicament. Today, the newspaper remarked, the energy supply of the country is facing a problem and the production of electricity at atomic and hydroelectric stations is lagging behind planned levels. Chernobyl had played a part in the deficit, but was hardly responsible for the dilemma in its entirety. After all, unit 1 of the nuclear plant was already back in service (it came on-line on October 1, 1986, see Chapter 6), and efforts were being made to resurrect units 2 and 3. Thermal electric stations had also compensated largely for the shortfall.[2] So where did the problems lie, in the view of *Izvestiya*?

First, construction work had fallen behind schedule not only at the three Ukrainian nuclear plants, but also at the water-driven boilers of the thermal power stations in Vinnytsya, Lviv and Kharkiv Oblasts. The building of new units at Kalinin nuclear power plant, and at Berezovsky and Gusinoozersky hydroelectric stations was being carried out "at a low tempo." Much of the responsibility for the slow pace, in the view of the editorial, lay with the Ministry for Energy Equipment of the USSR, which had not provided the workers with various equipment and parts, but the energy workers themselves were also to blame. Plans were corrected and amended in midstream. Capacities were not being used fully at plants in service, such as the Krasnoyarsk hydroelectric station, where over 40% of capacity was not being utilized.

Energy was being overused, according to the newspaper. In particular, workers in light industry, chemicals and oil machine construction were guilty of wasting energy supplies. Some factories had not warmed their premises in time, necessitating heavy use of heating during work hours. Others kept all the lights on during lunch times. The editorial asked why some of the peak period work was not being moved to evenings and weekends to preserve power.

Two days after this editorial appeared, the Central Committee of the CPSU held a meeting in Moscow to discuss preparations for the coming winter and ways of ensuring that the national economy received "reliable" supplies of energy. The main speakers at the session were Egor Ligachev and Vladimir Dolgikh. It was stated that many power stations were late with their repairs and that "extraordinary" measures were required for the winter period.[3] The impression one receives is that Chernobyl had sparked an unanticipated and serious energy deficit. The tardiness of repairs cited by the CC CPSU, for example, was a direct result of the accident, when the normal repair period was aborted so that other power stations could make up the shortfall. Despite the other factors mentioned, it was the accident that rendered the situation not merely serious, but acute.

Western media sources that adopted this somewhat natural conclusion, however, were severely admonished by the Soviet side. On October 28, for example, Soviet Television decided to take to task the Deutsches Welles radio station. The extreme sensitivity of the authorities on the question becomes evident from the nature of the program. Deutsches Welles had described some of the measures taken to "prevent a catastrophe" in Soviet electricity supply during the cold months. "Welt," stated the Soviet television program in response, "means world, and the world is not without good people." But what, it asked, evoked such "touching care."? "What has brought about such weeping?" Was it an awkward attempt to frighten the Soviet people? No, said the program, according to Deutsches Welles, it was all a result of the Chernobyl disaster.

Yet, it continued, the German radio station had avoided simple arithmetic. Only 4 million kilowatts of power had been removed from the country's energy supply out of a total capacity of over 300 million kilowatts. Is it justified to speak of an energy crisis in the entire country? "Of course not." Some 90 million kilowatts worth of equipment had been repaired and fuel supplies at electrostations "are significantly greater than last year." The country had a reserve of 2.7 million tons of coal. To accelerate the introduction of new energy structures, stated the program, about 8,000 qualified specialists from other construction ministries were being sent to power stations.

But what about the West?

A homeless man, jobless as a rule, sleeping on sewer gratings, the only accessible source of heat. Tell me, in what place in the world is this a symbol? Everyone knows, not ours, not Soviet.

Thus the program replied in fury to the report about the power crisis. Yet Deutsches Welles had only taken official Soviet sources to their natural conclusion. The power supply appears to be an issue on which the Soviets will give no quarter. In October 1987, a Soviet official was asked about the dimly lit Moscow streets and the fact that the cars in the streets used not their headlights but only their parking lights. Was there a power shortage in the capital? Not at all, he responded. But surely, he was asked, there was a power problem in 1986, after Chernobyl? Again he declared that there had been no such event. It did not exist, he stated, adamantly.[4]

A different picture began to emerge from official sources. One referred to the "serious electricity shortage" faced by the USSR "this winter" because of a variety of problems such as accidents, overexpenditure of electricity, poorly qualified personnel and inadequate organization. The Chernobyl nuclear accident and a summer drought were said to be contributory factors, and to compensate for the shortfall, coal-fired power stations had to burn far more fuel than planned (but what about the extra reserves of coal mentioned on Soviet television?). Accidents had resulted in a failure to supply over 500 million kilowatt hours of electricity in the first half of 1986, and many of these accidents were related directly to errors by personnel. "The picture has not changed in the second half of the year."[5]

A serious violation was exceeding the permitted capacity of electricity during peak hours. This could lead to the automatic exclusion of a large number of enterprises severely damaging work discipline and throwing the entire power system into turmoil. Such "violations" had reportedly occurred in the Azerbaidzhan SSR, the Estonian SSR, Perm and Ulyanovsk Oblasts in the RSFSR, and in the Chuvash ASSR. Factories using too much energy included the huge Zhdanov metallurgical combine, the Volgograd aluminum and oil-processing factories, and the Zavolzhsky motor factory.[6]

What about new power supplies being started up? There were some grounds for optimism. On October 11, it was announced that a third unit at the Zuyevsky thermal electric station near Kiev had been built in record time. It was 300 megawatts in size and would provide the Ukrainian power grid with an extra 200 million kilowatt hours by the end of 1986, thus compensating for the losses incurred after the

Chernobyl accident.[7] The report did not suggest that there might be other contributory causes to the shortfall: Chernobyl was responsible.

North of the border, in Gomel Oblast of Belorussia, it was stated early in 1987 that the industrial centers of that oblast had been experiencing shortages of electric power because of the Chernobyl accident. Therefore, it was necessary to bring on-stream a second unit at the Gomel thermal electric station at the end of 1986. Together with the two units at the Chernobyl plant in operation by November 1986, Gomel thermal station's No. 2 unit had "virtually solved" the problems of a steady supply of energy for the industrial center in Belorussian Polissya.[8] But despite this forthright announcement, the event had not happened.

This was confirmed by a report in the spring of 1987. At that time, it was stated that the second unit "was supposed" to start up at the end of 1986, but the suppliers had let down the builders. Half of the pipe joints prepared by the Belgorod factory of energy machine-building had arrived at the site with defects. Consequently, the city of Gomel was feeling the pinch of a power shortage.[9]

The key problem lay with bringing into operation the three new Ukrainian reactors in 1986. Of the three, Zaporizhzhya-3 was considered the most important and perhaps the only reactor that was certain to come on-stream on schedule. At Zaporizhzhya, two reactors had been constructed in a period of less than five years, and it had been announced that henceforth, the remaining four reactors would begin to work at annual intervals. The deadline for the third reactor was June 1986, at which time, it was confidently expected, an appropriate announcement would be made. This did not take place.

What had gone wrong? The factors involved were fairly typical of an average Soviet reactor, but evidently were unique to Zaporizhzhya, which has earned a place as a model for the rest of the Soviet VVER-type reactor construction. The Director of the Building Department at Zaporizhzhya, R. Khenokha, commented that in his view, the timetable for output in June 1986 was simply unrealistic. It could, nonetheless, have been achieved by September, but for a certain change of attitude after Chernobyl. People began to think, he stated, that it was neither feasible nor obligatory to adhere to such drastic construction timetables.[10] This is a somewhat strange attitude. The chaotic building schedule that had occurred before the accident, while not directly responsible for the series of events, had led to the demoralization of the workforce. Yet here was a building director who appeared to think that there was no reason to change the general outlook.

Other reasons also lay behind the delays at Zaporizhzhya. The factory supplying special construction materials for the reactor's covering, for example, was said to be two months behind schedule. The Ministry of Instruments of the USSR had provided computing equipment that turned out to be defective and a further two months were taken up in removing these defects. The Tashkent production association "Sredazelektroapparat" was supposed to deliver the control panels for the reactor more than 18 months before the commencement of construction work, but in fact had not sent them until January 1986, only five months before the reactor was due to come into service. The Atommash factory in Volgodonsk was also reportedly late in providing necessary equipment.

The leaders of the nuclear plant's building department had not provided their workers with technical documentation for the construction work ahead of time. The "Atomteploelektroproekt" even provided the layout plan for reactor 3 after building work had already begun. This sort of delay, according to the account, had not perturbed the building department unduly. Altogether, the list of problems at Zaporizhzhya-3 was said to make up 17 pages of text, and the result was the failure to start the reactor on schedule. Similar difficulties, albeit on a lesser scale, were said to be affecting work on the construction of Zaporizhzhya-4, which was taking place simultaneously. Thus the energy workers at Zaporizhzhya were accused of "being deficient" by 1 billion "atomic" kilowatt hours.

The delay in startup was having a direct impact on energy supply, witness the following comment from G. Dudnik, Chief Engineer of the Dnieper Energy System:

All enterprises under our jurisdiction—thermal and hydroelectric power stations—have prepared for work in winter conditions. Capital and average repairs have been completed. All the same, the energy deficit is serious. At times of morning and evening peak loads we have been forced to restrict even the metallurgists. In the entire republic, many users are on a curtailed ration situation. And now, when many enterprises have transferred to 2-shift work, they need additional energy. If the next block of Zaporozhye [Zaporizhzhya] atomic energy station was in operation, then the level of tension in the zone's facilities would be halved. Now once again this has not occurred....The real price of every kilowatt is being raised over and over again.[11]

The seriousness of the situation at Zaporizhzhya was demonstrated by the reaction of the Politburo of the CC CPSU. On October 11-12, it dispatched Candidate Member Vladimir Dolgikh to Enerhodar, the city for plant operatives of the Zaporizhzhya station. He was accompanied by Secretary of the Central Committee of the Communist Party of Ukraine, Borys Kachura, the member of the Ukrainian Politburo responsible for energy and heavy industrial affairs in the republic. At the station, a meeting was held, in which participated Directors and leaders of Union ministries and enterprises. The topic was the delay in the startup of Zaporizhzhya-3. There seems to have been an emphasis on the need to train more specialists for the nuclear energy sphere, but also many questions were apparently covered concerning the general work situation of the collectives, the housing and communal services at Enerhodar and others.[12] The Zaporizhzhya-3 reactor finally came on-line on December 11, 1986, six months behind schedule, but nevertheless still only 17 months after Zaporizhzhya-2.[13]

By the spring of 1987, early reports suggested that the worst was over. In January and February 1987, the country's electric stations produced 285.4 billion kilowatt hours of electricity, meeting the planned requirements. Because of the relatively favorable weather patterns, with the exception of Central Asia, there were no reported shortfalls of electricity at enterprises and factories. The Central Asian situation was affected less by Chernobyl than by the low water supplies at hydroelectric stations and by the fact that thermal electric stations in Uzbekistan were not working to their full capacity. Nevertheless, some industrial firms, especially in Ukraine, were said to be exceeding their electricity quotas during peak hours—for example, in Dnipropetrovsk and Zaporizhzhya.[14]

By April, the situation in the power industry was again declared to be "strained." The winter stocks at thermal electric power stations had been used up, and they were now short of fuel. An imbalance had occurred in the USSR's fuel-energy system. Because of considerable failures in the delivery of coal and fuel oil, the situation had deteriorated in Ukraine, Kazakhstan, Uzbekistan, Georgia, Armenia, the North Caucasus, the Buryat ASSR, Amur Oblast and Krasnoyarsk region of the RSFSR. Again, the main problem was declared to be overexpenditure of electricity, particularly in Ukraine where agricultural consumers had exceeded their quotas by 350 million kilowatt hours. Various ministries were taken to task for their lack of economy in electricity usage, notably the Ministry of Railways of the USSR, the Ministry of Ferrous Metallurgy of the USSR, and the Ministry of Petroleum Refining of the USSR.[15]

Thus Chernobyl brought about a short-term electricity crisis in the USSR that lasted into the spring of 1987. Other factors contributed directly to the dilemma, but had Chernobyl not occurred, it is safe to say that the situation could not have been labeled a crisis. Winter preparations are always a source of concern to the Soviet authorities and solicit a number of editorials in the Soviet press. In 1986-87, there was good reason to worry. There was, however, another area of concern, namely Eastern Europe, where at least two countries—Hungary and Czechoslovakia—were affected directly by the loss of the four Chernobyl reactors, and others, such as Romania, Poland and Bulgaria, also played a part because of a power shortage as a result of other factors.

Turning first to Hungary, the country imports over 30% of its electricity supply from the unified power grid of the CMEA countries, predominantly from the USSR. It is believed that Chernobyl-4 was providing power to Hungary at the time of the experiment that preceded the accident. In late October, the Hungarian news agency reported that Soviet electricity imports had "fallen slightly" in 1986 as a result of Chernobyl. Supplies to Hungary had reportedly been cut by some 500 million kilowatt hours from 10.5 to 10.0 million kilowatt hours, a drop of slightly less than 5%. Moscow was said to be making up for the shortfall with additional supplies of natural gas, but evidently it did not compensate for all of it.[16] The problem may have been that the small shortfall occurred all at once late in the year, causing Hungary to shut down some factories during a cold spell in September.[17]

In November, at the 42nd Session of the CMEA countries in Bucharest, the Hungarian Premier, Gyorgy Lazar, spoke about the state of affairs in his country's power supply. He maintained that there were "occasional disorders" in the unified system, and that "certain partners" had dipped into the Hungarian supply. Such irregularities in the most recent period, he stated, "have necessitated immediate, at times drastic, government measures, causing extremely serious harm to the Hungarian economy."[18] A Western authority points out that the problem was caused by Chernobyl, but that the energy was available. The difficulty was that the power networks needed to be redeveloped because they were unable to deliver electricity to the Mir power grid of Eastern Europe.[19]

The Chernobyl explosion also resulted in a drop of Soviet electricity supplies to Czechoslovakia. Vlastimil Ehrenberger, the Minister of Fuel and Power in Czechoslovakia stated in mid-November 1986 that the country had to raise its exports of electric power in return for deliveries of oil. The combination of increased exports and Soviet cuts meant that Czechoslovakia had to produce over 2,000 million kilowatt hours of electricity more than planned for 1986. The total cut was said to be 1,200 million kilowatt hours, but the lower river levels in the country, as in the USSR itself, had brought the total shortfall to 1,700 million kilowatt hours. Once again, the Soviet Union had reportedly increased its exports of natural gas to make up for the electric power deficit.[20]

Turning again to the USSR, as the designed frequency of the power grid fell down from 50 to 49 hertz, threatening to shut down the motors in numerous factories, the Chernobyl accident brought about another unforeseen consequence, namely the question of the future of the RBMK reactor itself.

The Debate Over the RBMK

As noted, the disaster led to the subsequent shutdown and retrofitting of all 14 Soviet RBMK reactors in the summer and fall of 1986. Subsequently, the two Chernobyl reactors were restarted in October and November 1986, and many Western observers felt that the RBMK building program would continue. In Vienna, the Soviet report emphasized the overwhelming significance of human mistakes in causing the accident. The Soviet delegation received widespread praise for its report and for its relative openness and frankness about the accident. In turn, however, by dealing so closely with the IAEA, future cooperation with the UN body necessitated regular IAEA visits to Soviet reactors. Also, nothing could hide the fact that had the RBMK reactor been better designed, the accident could not have occurred. The Soviets operated under the assumption that their operators would avoid unsafe regimes, rather than from the standpoint that unsafe regimes should be made impossible to bring about. The consequence of both the above was a belated debate on the future of the RBMK, which will have a major impact as far as the longterm economic consequences of the accident are concerned.

The RBMK, the first and an exclusively Soviet type of reactor, still constitutes the majority of nuclear power capacity in the Soviet Union. The first generation of RBMKs began at Leningrad nuclear power plant, and other plants are in various stage of construction at Chernobyl, Ignalina (Lithuania), Kursk, Smolensk and work was in progress on new units at Kostroma in central Russia. The current building program intended to raise the number of units from 14 to 21 reactors, either of the type RBMK-1000 or RBMK-1500. The latter design was to have been built only at Ignalina and Kostroma stations. Ultimately 29 reactors were planned. Before Chernobyl, Soviet reports had invariably emphasized that the future nuclear energy program to the year 2000 would rely on both RBMK and VVER units. There was no indication that one would eventually predominate over the other.

By January 1987, there were signs that the Soviet outlook was changing. The changes coincided with, but may not have been connected with a visit by IAEA Director-General, Hans Blix, to the Chernobyl plant in the middle of the month. Almost simultaneously, a Soviet Deputy Director of the IAEA, Leonid Konstantinov, informed Western reporters in Moscow that no new RBMKs would be built once the current program of construction had been completed. A few days later, an IAEA spokesperson, James Dagleish, stated that the Soviet Union was undecided over whether it should abandon plans to build future RBMKs. During the Blix visit, it had been made clear that the construction work already under way on 6 RBMKs would be completed. Evidently the six reactors in question were Chernobyl-5, Chernobyl-6, Smolensk-3, Smolensk-4, Kostroma-1 and Kostroma-2.[21]

By early February 1987, the new plans had been confirmed, with the exception that two new RBMKs had been added to the six above, namely Chernobyl-3 and Ignalina-2.[22] The addition of Chernobyl-3 was somewhat surprising in that for the most part, Soviet sources have included it automatically in their overall capacity total. Although the reactor was shut down after the destruction of its neighbor, No. 4, with which it shared a building, there was no physical damage to Chernobyl-3, and it was felt that the reactor would be back in service by June 1987. Although the timetable was subsequently changed to December 1987, the Soviets always acted as though it was never really out of commission.

The Ignalina plant is a different matter. It has been regarded as dangerous by the Swedish authorities since the first 1,500 megawatt unit came into service in December 1983. Unit two was scheduled for 1986, but it did not come into service during that year, leading to rumors that the entire station was to be abandoned for safety reasons. The Swedish press played a key role in carrying these reports. For example, on April 12, 1987, the Swedish daily, *Upsala nya tidning*, purported to quote an official of the Ministry of Atomic Power Engineering of the USSR, Evgenii Larin, as saying that the plans to expand the station (presumably the reference was to units 3 and 4) had been scrapped because the station's technology was obsolete.

On the next day, Radio Stockholm broadcasted a story from its Moscow correspondent, Kjezl-Albin Abrahamsson. The latter confirmed that the Ignalina-4 reactor would not be built as planned, partly because of objections to the plant within Lithuania itself. He also noted that there were rumors in Moscow that Smolensk-4 would not be built either. The Swedish accounts stressed safety factors. The Swedish Defense Research Institute has reportedly monitored six small radioactive leaks from the plant since 1983, and Sweden's energy experts have maintained that because of its 1,500 megawatt capacity, Ignalina is actually more dangerous than Chernobyl itself.[23]

There were also some doubts about Ignalina-2 because Radio Moscow had announced on February 19, 1987 that its start-up was imminent at the end of that month, and then nothing further had been reported. These doubts were put to rest only in mid-August 1987, when the Soviets declared that unit 2 was now in service. The six-month delay was attributed to the implementation of additional safety measures: changes guaranteeing the safe operation of the reactor's fuel channels and the means to stop an atomic reaction (perhaps the reduction of the time required to shut down the reactor). Similar changes were envisaged for Ignalina-1.[24] Ignalina-2 was the first post-Chernobyl RBMK to be put into service and it went very much against the prevailing trend. As far as the RBMK program was concerned, the situation had become precarious.

In March 1987, Valerii Legasov, First Deputy Director of the Kurchatov Institute at the Soviet Academy of Sciences, and the leader of the 1986 Soviet delegation to the IAEA in Vienna, was interviewed by Novosti Press Agency. Legasov maintained that when the Atommash factory for manufacturing equipment for the VVER reactor was built in Volgodonsk, it was planned that the VVER would be the exclusive reactor under construction in the future. Now that this industrial giant had begun to make its mark, decisions were made to complete the erection of 8 more uranium-graphite reactors, and then to switch over completely to VVERs.[25] It should be pointed out that despite the publication of numerous items about Atommash, especially in the Soviet daily newspaper, *Sotsialisticheskaya industriya*, there had been no indication that such a change was likely. It was thus either a well-kept secret, or Legasov was thinking retroactively!

However, Legasov's comments were supported by A.M. Petrosyants, Chairman of the USSR State Committee for the Utilization of Atomic Energy. Petrosyants provided a very cautious account of the situation. The new safety measures implemented at Soviet RBMKs, he stated, made it possible to continue their safe operation, and power generation based on the RBMK "may develop further." But the only reason for the delayed dominance of the VVER was the need to manufacture the huge steel vessels and containment domes for this type of reactor. Atommash had resolved these twin problems and therefore the future development of nuclear power generation in the USSR was to be based on VVERs, a decision that had been taken before the Chernobyl accident.[26]

While Soviet scientists were asserting that they were simply putting into operation old plans, the reality appeared otherwise, even to the most objective observer. Only a day before Petrosyants' interview appeared in the Russian press, he had announced over Radio Moscow that the construction of units 5 and 6 at Chernobyl "will not continue."[27] The statement contradicted every previous statement made about the future of these reactors,[28] and even if one accepts that a longterm strategy had existed beforehand, this does not explain why plans were changed so dramatically in midstream. Chernobyl-5, for example, had almost been completed at the time of the accident. Moreover, if the radiation situation was the chief concern, then why were plans still going ahead to restart Chernobyl-3 in 1987?

As far as Chernobyl-5 and Chernobyl-6 were concerned, there may have been factors other than safety involved (see Chapter 6). But by October 1987, it had been revealed that the Kostroma-1500 reactors were to be converted to VVER-1000s, "for economic reasons."[29] Thus of the original eight reactors to be completed as planned, only the two Smolensk reactors and Ignalina-2 appear to have been unaffected at the time of writing. Four had been abandoned, and Chernobyl-3 appeared to be in some doubt, despite official optimism. Thus by the end of 1987, the RBMK program, despite alleged safety improvements, was in a shambles. The evidence suggests that Chernobyl forced the Soviets to rethink their entire nuclear energy strategy. Having said that, it must then be acknowledged that the proposed rate of expansion of the nuclear power capacities in the country did not change as a result of the Chernobyl disaster.

Initial Impact on the Soviet Energy Program

Along with the new emphasis on the VVER-type reactor, the Soviet nuclear energy program, at first, was to continue with unabated vigor. This was a somewhat surprising development in view of the slowdown in nuclear reactor construction elsewhere in the world as a result of the Chernobyl accident. However, it is confirmed by various official Soviet sources during the 18 months following the accident. In the spring of 1987, the Soviet outlook on nuclear power and the energy situation of the country generally was described by Yurii Bublyk, head of the Laboratory of All-Union Scientific Research Institute of Fuel Energy Problems of Gosplan USSR in an interview in the newspaper *Sovetskaya Rossiya*.[30]

Bublyk noted that whereas 75% of fuel consumers live to the west of the Urals, 80% of the country's resources are to be found in the east. Fully 40% of all the goods transported from east to west consist of fuel. What is the solution? We should try, he stated, to transfer the maximum possible number of energy works to the east, while at the same time raising the efficiency of energy consumption in the European part of the country. "But our capabilities are not infinite." The use of hydroresources in European USSR is already twice as high as in other regions. The transportation of coal from Siberia to the west is "a great waste of energy" and does not in any case resolve the problem of guaranteed heat supplies for the national economy. Thus "there is no sound alternative to the accelerated development of nuclear energy."

Bublyk's comments would have satisfied the Politburo of the CC CPSU. In April 1987, the Politburo held a meeting on "Raising the technical level and the reliability of energy equipment, the quality of its exploitation." One of the main speakers was Secretary of the CC CPSU, L.N. Zaikov. He stated plainly that the question of fuel-energy resources in the country was not proceeding smoothly. In "energetics," stagnation had occurred, there were examples of "lounging around." One of the chief problems was that power units were not being brought into operation on schedule. This was partly a technical problem in that Soviet equipment was often poorly made and inferior to foreign models, but, in Zaikov's view, it was the bureaucratic approach to energy questions that was the main issue. The schedules, therefore, had to be strictly maintained if the country was to meet its energy requirements.[31]

The dimensions of Soviet reactor unit construction are staggering. There were 54 reactors in operation in 1987, with a total capacity of 34,469 megawatts. A further 29 were under construction, with a capacity of 31,000 megawatts, while 27 were planned, at a total of 28,000 megawatts.[32] Aside from the question of reactor types therefore, one can say that shortly after the accident occurred, the USSR was about one-third of the way through its current program. In terms of kilowatt hours produced, the goal was to raise output from 170 to 390 billion between 1985 and 1990. The program's goals are well known and do not require elaboration here. What is of concern is whether Chernobyl had any immediate impact on the expansion program.

At the meeting of the CMEA countries in Bucharest in November 1986, Soviet Premier, Nikolai Ryzhkov, emphasized that there was to be no slowdown to either the East European or the Soviet programs for nuclear power development. A document had been drawn up concerning expansion up to the year 2000. It stated that in the CMEA countries outside the USSR, total capacity was to be raised from 8,000 megawatts in 1986 to 50,000 megawatts by the end of the century. In the USSR, it was anticipated that the rise in capacity would be 500-600% and that the proportion of nuclear-generated electricity in the country would rise correspondingly from 11 to 30%. In addition, said Ryzhkov, nuclear heat-supply stations would be built that would enable considerable savings on valuable and scarce organic fuel gas and oil.[33]

The policy had been developed following what was described as a "painstaking analysis" in the light of the Chernobyl accident. According to Ryzhkov, the tragedy had not only not ended prospects for nuclear power engineering cooperation in the CMEA countries, but

> by focusing attention on ways of achieving greater safety, it has increased its importance as the sole source that can guarantee reliable power supplies in the future.

According to Andronik Petrosyants, nuclear energy would increase in importance as the years pass. By 1990, if the plan targets were met, it would grow to 21% of all the electricity produced, and the volume of its increase over 1985 would be 70%. In the first stage of the USSR Energy Program therefore, the expenditure of organic fuel in the European part of the country was to be greatly reduced as a result of the construction and operation of nuclear power plants. In the second stage, 1990-2000, nuclear power was to make up the main portion of the growth in energy.[34]

The new spokesperson for the industry was Nikolai Lukonin, Minister of Atomic Power Engineering (a ministry established following the accident). Lukonin also emphasized constantly that the "strategic line" concerning nuclear power development in the USSR "despite Chernobyl, remains unchanged." Electricity generation was to be doubled at nuclear plants in 1990, as compared to 1985, he stated, and to be more than trebled by 1995. Lukonin reemphasized the plans at the IAEA meeting in Vienna in September-October 1987, for which he prepared a paper on the subject.[35]

Lukonin's deputy, Aleksandr Lapshin, spoke at a press conference in July 1987 that was dedicated to the 30th anniversary of the IAEA, an event that was marked by a number of articles in the Soviet press. He noted that by January 1987, the USSR was in third place in the world in terms of the volume of electricity produced at its 17 currently active nuclear plants. Electricity production, however, stood at only 11% of all Soviet electricity production and had thus not risen in proportion since 1985. Eleven new stations were being built, Lapshin announced, before repeating the ambitious goals of the program.[36]

It is evident therefore that despite the doubts cast on the viability of the RBMK reactor, emphasis was at first switched to the VVER and the program was to continue unabated. (As will be shown in Chapter 7, new developments have prevented this from occurring.) This is, on the face of it, one of the more remarkable consequences of Chernobyl. It is not to say, however, that the goals of the program could have been met. Indeed, one of the main stumbling blocks was the Atommash factory on which so much depended now that the RBMK program had been abandoned. Since the debacle that occurred at Atommash in 1983,[37] considerable attention has been focused on the reactor manufacturing giant. But it has always fallen behind proposed schedules as far as construction and installation work are concerned.

For example, in October 1986, it was noted that whereas the tempo of work is normally raised toward the end of the first year of a new five-year plan, at the Atommash factory, the reverse was the case. The complex of the third building at the plant was said to be the only one at which the plan of assimilation of capital investments had been completed by more than 50% in September 1986. The builders were reportedly working "at an extremely slow tempo," and it was urgent that the state of affairs at the construction site be changed "in a fundamental way."[38] This report was typical of many on the factory.

By the summer of 1987, Atommash had the capability of producing two VVER-1000 reactor complexes a year, as opposed to its anticipated schedule of four per year, rising to eight by 1990. Even then, some of the factory's customers could not accept equipment because they had fallen so far behind schedule in their building work at the plant sites. Two steam generators prepared for the Rostov nuclear power plant had been "sitting idle" at Atommash since 1984 because the nuclear plant had no space in which to store them. Similar problems arose at the

Crimea and Balakovo nuclear plants. Back in Volgodonsk, this equipment was being shunted around from one place to another, necessitating valuable work time and use of facilities.[39]

As a result of the undelivered products, a financial crisis had occurred. At the start of 1987, the Atommash collective appealed to the Ministry of Power and Electrification of the USSR for help. Deputy Minister V. Kondratenko, responding through the medium of the newspaper *Sotsialisticheskaya industriya*, stated that measures had been elaborated that would permit the guaranteed receipt of equipment at Rostov, Crimea and Balakovo nuclear power plants, commencing in April 1987. By late July, however, no measures had been undertaken, and the factory now had 12 steam generators—an entire year's output—in stock and with no prospect of delivering them to customers. The dilemma illustrates that it may not have been wise to place great confidence on operations at Atommash, as Soviet scientists such as Petrosyants and Legasov had evidently done. Work there has never lived up to expectations and the factory has stumbled from one crisis to another. Yet herein lies the future of the massive Soviet nuclear power expansion.

The Ukrainian Nuclear Energy Program

Chernobyl brought home to many observers the significance of the Ukrainian SSR in the Soviet nuclear energy program. Whereas the Soviet plan noted above envisages that the share of nuclear energy in electric power production will be 30% in the USSR as a whole by the year 2000, in Ukraine, it will rise to 60%. By January 1988, the Ukrainian industry had a total capacity of 11,880 megawatts, i.e., 33% higher than at the time of the Chernobyl disaster (see Table 1). However, the proportion of electricity produced by nuclear energy in Ukraine was by then about 22%, compared to an average of about 11% in the USSR as a whole.

In 1986, two of the three planned VVER-1000 reactor units came on-stream, namely Rovno-3 and Zaporizhzhya-3. A long delay occurred at Khmelnytsky-1 (see below), and at the first unit of the Crimean nuclear power plant, at which two VVER-1000s must be in service by 1990 if the plan is to be met. Also before 1990, two more units are planned at Zaporizhzhya and two at the South Ukrainian plant. Nuclear power and heating plants were being built at Odessa (see Chapter 7) and Kharkiv, relatively close to the centers of those cities. At the former plant, four 500 megawatt reactors were planned, while

Table 1   Ukrainian Reactors in Operation, January 1988

Unit	Reactor-type	Year
Chernoby-1	RBMK-1000	1978
Chernobyl-2	RBMK-1000	1979
Chernobyl-3	RBMK-1000	1981 (1987)
Rovno-1	VVER-440	1979
Rovno-2	VVER-440	1982
Rovno-3	VVER-1000	1986
South Ukraine-1	VVER-1000	1982
South Ukraine-2	VVER-1000	1985
Zaporizhzhya-1	VVER-1000	1984
Zaporizhzhya-2	VVER-1000	1985
Zaporizhzhya-3	VVER-1000	1986
Zaporizhzhya-4	VVER-1000	1987
Khmelnytsky-1	VVER-1000	1987
Total capacity	11,880 megawatts	

at the latter, the original schedule called for two 1,000 megawatt VVERs. A similar plant to be built within the vicinity of the city of Kiev was abandoned suddenly in October 1987, and the proposed station is to be converted into another thermal electric station, thereby negating one of the reasons for construction of the new stations, i.e., cutting down on organic fuel consumption.[40]

Putting the Ukrainian program into perspective, the longterm program to the year 2000, as revised because of and immediately after the Chernobyl accident, still foresaw a capacity expansion from the early 1988 total of 11,880 to approximately 29,100 megawatts or by 330%. This increase, although massive in scale, was slightly less than the overall rise of 500-600% anticipated for the USSR, but Ukraine already has the highest percentage of nuclear power plants within the Union. The actual total may have been higher, in that no clear plans had been stated for the ultimate size of the new plant being built at Chyhyryn in Cherkassy Oblast. We have assumed an average size of 4,000 megawatts, but several stations, such as Zaporizhzhya and South Ukraine (as also the original plans for Chernobyl) were scheduled to be over 6,000 megawatts in size.

Within the USSR, Ukraine at the end of 1987 had 20 out of 58 Soviet reactors, or more than one-third of the total. In terms of capacity, it possessed 31.6% of the all-Union total. Had the Chernobyl accident not occurred, that percentage would now be 34.5%, which is a more appropriate indicator of the role of the Ukrainian SSR within Soviet nuclear energy strategy. The Ukrainian program is the most important sector of the USSR's nuclear plans, and in 1986-87, there appeared to be no anticipated slowdowns.

After Chernobyl, the Ukrainian Writers' Union and various Western sources pointed out that various warnings had been issued in the Soviet press about the sloppy construction work, poor supplies and defective equipment that were being used to build nuclear plants throughout the USSR. Lyubov Kovalevska's article in the March 27, 1986 edition of *Literaturna Ukraina*, for which she was initially demoted within the Prypyat newspaper hierarchy and rebuked, was cited by many prominent Western newspapers as an ominous warning of what was to occur at the Chernobyl plant one month later. Although Soviet officials have stressed that the accident itself was a combination of almost inconceivable circumstances, they have by-and-large admitted that there have been faults with the way plants have been constructed in the Soviet Union.

Does this mean, then, that in the early aftermath of Chernobyl, there were changes in the way stations were being constructed? Was the quality control higher not only in light of Chernobyl, but in the era of Gorbachev's perestroika? Had the authorities accepted that for safety reasons there must be occasional slowdowns in schedules, that the nuclear industry cannot be readily adopted to crash plans? Again, we will focus on the Ukrainian case since it is the most relevant.

As will be shown in Chapter 7, significant changes had occurred in the nuclear energy sphere as far as making reactors safer and training is concerned. But these had not affected substantially either the pace of the program or construction work at existing reactor sites, at least until late 1987. In the former case, we have noted that because there was an electricity crisis in the USSR in the Fall of 1986, the first two Chernobyl reactors were brought back on-line before any of the preliminary work on the area had been completed. Unit 3 was still being decontaminated. The shift settlement of Zelenyi Mys (see Chapter 6) had not yet been built for the plant operatives. And above all, none of the proposed longterm improvements to the RBMK-1000s had been implemented, which was of particular importance at second-generation RBMK reactors such as Chernobyl-1 and Chernobyl-2. When economic needs required it, the reactors were put back into service regardless of safety factors. Chernobyl did not change the Soviet attitude in this respect.

What is even more disturbing is that some of the "horror stories" that appeared about building work on nuclear plants before April 1986 were finding their way back into the press. Two articles in *Radyanska Ukraina* in December 1986 described the chaotic state of affairs at the Crimean station.[41] At the Chernobyl Trial (described below) of plant operators, the in-camera session evidently discussed an accident that occurred very recently at the Rovno station, which appears to have had serious structural problems linked to the location of the building site.[42] Whether a major accident involving casualties occurred is not known (although it seems unlikely that there could have been any leakage of radiation). Whatever did happen, the affair was considered serious enough to preclude the presence of Western reporters, even during a period of glasnost.

But perhaps the most disturbing news was that about work on the Khmelnytsky nuclear power station in western Ukraine. The station is a CMEA enterprise in which the Soviet Union and Poland have the largest stake (there are some 2,700 Polish workers at the site), and Hungary and Czechoslovakia have also invested in the station. At its peak capacity, according to the original plans, Khmelnytsky station was to provide 4,000 megawatts of electrical capacity for Ukraine and its East European neighbors. Each country that has invested in the project is to receive electric power in proportion to the size of the investment. However, although the USSR has the largest investment, it is the significance of the plant in Poland's electricity supply that is most notable.

The Polish nuclear power program originally envisaged a total capacity of 7,860-9,860 megawatts by the year 2000.[43] But the Zarnowiec station in Gdansk province has been constantly delayed and the timetables revised so that in 1987, Poland still had no nuclear reactors. In an article in a Western source, Peter Wood has pointed out that the plans to put Zarnowiec-1 into operation by 1990 (the initial timetable was for operation in 1984) have been delayed for at least a year. He maintained that one reason was that the foundations of the reactor were cracking.[44] According to Wood, the Polish authorities had ignored a report from the Department of Geology at Gdansk University which stated that the station's foundations were not suitable for the massive edifice to be built, and today therefore the future of the plant is in jeopardy.[45]

The Polish authorities have acknowledged that there are problems at Zarnowiec. At a meeting held at the building site in April 1987, it was noted that there was a shortage of both workers and building equipment, particularly the deliveries of high-grade cement, and upgraded steel for the plant's turbines.[46] Later in the year, the newspaper *Zycie Warszawy* stated that construction work at Zarnowiec had fallen 2-5 months behind schedule, and that the first reactor would not be connected to the grid until the end of 1991.[47] Polish Minister of the Environment, Stefan Jarzebski, was quoted as saying that the pace of building at Zarnowiec had been affected by the increased attention to safety after Chernobyl and that he was "personally satisfied" with the slowdown in tempo.[48]

Whether for safety, equipment or geological reasons, the protraction of the timetable for building Zarnowiec also made it more imperative for the Soviet authorities to hasten work on the Khmelnytsky station over the border, where construction work had been continuing for almost a decade. Originally scheduled for the 1981-85 plan period, Khmelnytsky-1 was eliciting attention at the highest levels of the CC CPSU. In 1983 there was a major shakeup at the station following a scandal involving the theft of various items of equipment, mismanagement, "violations" in work and other factors.[49] By 1987, the situation there appeared to have become even worse.

In May of that year, it was revealed in a Ukrainian newspaper that the plan for using basic production funds had been fulfilled by only 11.6% in 1986. Fourteen designated objects of industrial construction had not been completed as scheduled. In turn, the building of social and cultural amenities was well behind schedule.[50] Various complexes that should have opened in 1986 were still unfinished. They included a commercial center, a clinic, a house of culture, a sports center, and a bread factory. In 1987, the situation had barely improved. In the first quarter of the year, the program for building-assembly work had been fulfilled by only 66%. On the reactor complex itself, plan fulfilment stood at 86% for the first four months of the year.[51]

As a result of the acute construction problems at the Khmelnytsky station, the Deputy Director of the Construction Department, N.G. Akhtyamov, was removed from his post after a meeting of the Khmelnytsky Oblast Party Committee on February 20, 1987. At that time, the Director of the department, E.O. Bazhenov, was severely censured and threatened with more serious punishment if matters did not improve dramatically. In March 1987, five leading officials at the plant were given reprimands on their party cards, including the Chief Engineer, A.P. Selykhov. More reproofs followed, but two months then elapsed without any noticeable improvements.

The main problems included changes in planning decisions—which were occurring "like an avalanche"—and a failure to provide the needed technical documentation for building-assembly work in good time. In addition, thefts had continued from at least 1983 onward. In 1986, over 45,000 rubles' worth of goods were stolen from the building fund, partly at the instigation of railroad workers whom, it is said, were inviting the station's authorities to help themselves to the equipment that had been brought to Netishyn by rail.

The article pointed out that in the very month the Khmelnytsky station's first reactor had been scheduled (but failed) to come into service—December 1986—Bazhenov's building department was awarded the Red Banner and Scroll of Honor "for achieving the highest results in work."

In July 1987, a second major report on the Khmelnytsky nuclear plant appeared in the newspaper for Ukrainian youth, *Molod Ukrainy*. An account written by Olena Talayeva was sarcastic in tone and reminiscent of Kovalevska's approach in her original attack on the state of affairs at Chernobyl in *Literaturna Ukraina*. She described the December 1986 awards to Bazhenov and his colleagues as "empty phrases" and "pompousness." In her view, such things were meaningless.[52]

She also described a bureaucratic foul-up that had taken place in establishing which particular department had jurisdiction over a broken crane that was badly needed for building work. On the one hand, she wrote, the crane was being used by the Oliinyk brigade that was building new housing settlements in the area under the supervision of the Southern Energy Assembly. But on the other, the crane was actually owned by the Department of Mechanization at the nuclear plant itself, headed by M.F. Biryukov. To obtain permission to use the crane, the Oliinyk brigade had to kick its heels for over a month. The appropriate authority for resolving such questions, the State City Technical Inspection of the Ukrainian SSR, located in Kiev, had to deal with "several gross defects" in other parts of the plant, and these defects took priority.

The author's main premise was that the delays in building the first reactor unit and residences for plant operatives were a result both of bureaucratic inertia—Biryukov was portrayed as an obstructive man who cared little about the plight of the bricklayers who needed the crane—and the slow introduction of "new economic methods" into the building projects. The bricklayers had written to the editors of *Molod Ukrainy*, a newspaper that had featured the construction of new reactors quite frequently over the past few months,[53] inviting them to come to Netishyn and explain the meaning of perestroika to the plant's administration. Talayeva commented that the builders were growing accustomed to terms such as *dovkhobud* (long drawn out building projects) and "one does not have to be a genius to understand why."

The author also revealed that sociology students at Kuibyshev State University had recently carried out a survey of 400 workers at the Khmelnytsky station. Concerning labor productivity, one in three respondents had said that they considered their current performances inadequate, while 51% of those questioned believed that the workrate could be raised by 150% if it were tied directly to the amount of construction completed. The latter was perhaps inevitable in view of the fact that a 6-7 year building project had actually taken over 9.5 years without any tangible results!

Talayeva wrote that after "mountains" of memoranda had failed to uncover the real reason for the backward state of construction work, the Oliinyk brigade was allowed to work under "new conditions," by which one assumes that work and bonuses were distributed according to the amount of building work actually completed and that the enterprise itself was given control over its own accounting and profit distribution. But almost immediately, the new methods had encountered "the old counteractive forces: irresponsibility and mismanagement." To date, the "battle" had been an unsuccessful and uphill affair, wrote the author. In brief, even the introduction of economic reform into building work had not fundamentally altered the situation at Netishyn.

The above situation appears to be a typical one at Ukraine's nuclear power plant constructions and it is a state of affairs that has not been affected by Chernobyl. Khmelnytsky nuclear plant was actually the most delayed of all the reactors under construction, but the authorities have never managed to adhere to the original schedules and quality control has remained at a low level. On December 23, 1987, the Kiev newspaper, *Radyanska Ukraina*, announced that the Khmelnytsky-1 reactor had finally been put into service, almost 10 years after building work had begun on its foundations.

In this author's view, two factors are evident in the Ukrainian program. First, the ambitious goals to the year 2000, even before the late-1987 and early-1988 anti-nuclear power movement developed in Ukraine, could not possibly have been met at the current workrates and with the current problems related to safety concerns. As noted, the Kiev nuclear power and heating plant had been abandoned, ostensibly because of the need to create "new and wider safety barriers after Chernobyl." At the same time, the Crimean station, originally scheduled to be completed by 1990 will actually have only one of its reactors onstream by this date.[54]

Second, nuclear power is becoming undeniably one of the principal and most important industries in the Ukrainian SSR. In this latter development, the East European link is a major factor, but it also cannot be denied that nuclear power with its "guaranteed" supply of electricity for major industries is perceived as a much more reliable energy source than the declining Ukrainian coal industry, and even oil because the main reserves of the latter are located in the eastern regions of the USSR. The Zaporizhzhya station is serving as the model for a series of VVERs being constructed according to a unified design with identical reactors, turbines and other parts. To reiterate: Chernobyl did not change the basic outlook of the Soviet authorities on nuclear energy development, although as will be shown below (Chapter 7), public opposition has seriously threatened the expansion plans. Today, Ukraine remains a central and vital part of the nuclear program.

Political Consequences

Chernobyl was the first of several accidents to occur in the USSR in 1986 for which a Government Commission was appointed to deal with the consequences. The decision to appoint such a Commission may have been makeshift in that the full scope of the accident may not have been known during the first hours when the Commission was appointed, it nevertheless remained in operation until its main functions were usurped by the establishment of the Kombinat production association in November 1986 (see Chapter 5). Similar Commissions were later set up to deal with two Ukrainian coal mining accidents in December 1986 and May 1987, for two train crashes in Ukraine and in Rostov Oblast of the RSFSR, in November 1986 and August 1987, and for the collision of two vessels on the Black Sea in the late summer of 1986. Following the latter accident, a trial took place which may have served as the precedent for the trial of Chernobyl plant operators in July 1987.

After the Government Commission had been created to deal with the consequences of the Chernobyl disaster, the Soviet authorities came under fire internationally for what was perceived as a failure to take actions appropriate to a major disaster that was affecting countries outside the USSR. Within the Soviet Union itself, two distinct schools of thought have emerged on this question.

First, there is what might be termed the "glasnost" school that remains heavily critical of Soviet actions immediately after the accident, and continues to question followup policies within the zone, such as the rapid repopulation of evacuated villages (Chapter 6). The school is epitomized by journalists such as *Izvestiya*'s Stanislav Kondrashov and Andrei Pralnikov, by *Pravda*'s Vladimir Gubarev, and by the weekly newspapers *Moscow News*, *Ogonyok*, and *Nedelya*.

Second, there is also a powerful "revisionist" school that seeks basically to minimize the impact of the accident, particularly outside the Soviet Union, and refuses to acknowledge criticisms of post-accident operations. It is supported most vigorously by prominent Soviet scientists such as Valerii Legasov and Leonid Ilyin and it has used the entire Chernobyl episode to illustrate two main points: namely, that the event was sensationalized by the Western media, which blew matters all out of proportion by greatly exaggerating the casualties; and that Chernobyl proved once and for all that the only sensible policy for world leaders to follow is Gorbachev's scheme to remove all nuclear weapons from the face of the earth by the year 2000. Although both points contain more than an element of truth, both are also irrelevant to the study and the nature of the disaster itself.

The crux of the debate between the two schools of thought within the USSR centers on the following "failures":

1. The Soviet authorities failed to report the accident for 48 hours. The Moscow press carried only a two-sentence announcement from the Council of Ministers on April 30, 1986, four days after the event. According to Kondrashov, however, the Moscow newspapers were immediately made aware of what had occurred, and at *Izvestiya*, a furious debate had ensued over whether a story about Chernobyl should be printed on the following day. While the idea was rejected narrowly, he is in no doubt that where the Chernobyl disaster to occur again, a story would appear immediately, a point borne out by the relatively speedy coverage of subsequent accidents.[55]

2. The local party authorities encompassed by the former Prypyat City Party Committee, under the direct jurisdiction of the Kiev Oblast Party Committee, had underestimated the scale of the disaster or had believed that it could be "contained" locally. Therefore they had neither taken the necessary measures nor informed Moscow of the enormity of what had occurred (this is contradicted somewhat by Kondrashov's statement above).[56]

3. No warnings had been issued to the population in the area before it was informed on the following afternoon that an evacuation would be necessary. Up to the moment of the evacuation of Prypyat, people were walking outside in the streets. We have described elsewhere how children had attended Saturday school only hours later, and how farmers were grazing cattle in ditches contaminated by

radioactive fallout. The first health warning to the population of the 30-kilometer zone came only on May 5, 1986.

4. The Ukrainian party hierarchy, especially the Kiev Oblast Party Committee under Hryhorii Revenko, and ultimately First Party Secretary Volodymyr Shcherbytsky since he was also located in Kiev, had decided finally to evacuate only a small area of approximately 10 kilometers around the damaged plant. Not until a delegation arrived from Moscow, headed by Egor Ligachev and Nikolai Ryzhkov, on May 2, 1986, was the evacuation area extended to a more realistic 30 kilometers.

5. While individual firemen and military helicopter pilots had shown great fortitude and self-sacrifice in battling the uncontrolled reactor and a graphite fire that lasted for weeks afterward, others had "failed a major test." These included in particular Viktor Bryukhanov, the plant Director, and Mykola Fomin, its Chief Engineer, neither of whom should have permitted the experiment on the reactor to take place. Bryukhanov, as will be shown below, was accused of deliberately misleading local authorities about the scale of the accident and the actual levels of radiation in the area. Some 177 local party members had fled from the scene after the accident and many could not be located for weeks. By August 1987, 10% of Chernobyl plant's personnel had had to be replaced.[57]

6. The operators who had carried out the experiment and the ministries and state committees in charge of the Chernobyl plant were also held responsible for the almost incredible series of errors that occurred on April 25-26. It was the ministries rather than the party hierarchy that suffered the main repercussions of Chernobyl.

The political fallout from Chernobyl was determined by the above points. In addition, it could not be separated from the prevailing political and economic climate within the USSR, which had experienced the first year under new General Secretary of the CC CPSU, Mikhail Gorbachev. It became clear to many Western observers that whether or not the authorities in Moscow had been kept in the dark about the nature of the accident, clearly Shcherbytsky's Communist Party of Ukraine had to appear to be the guilty party.

Volodymyr Shcherbytsky, by the spring of 1986, had already become something of an anomaly within the CC CPSU Politburo. He and his contemporary, the Kazakhstan First Party Secretary, D. Kunaev (along with Andrei Gromyko who had been given the ceremonial position of President by Gorbachev), were perceived as being of the "old school," i.e., both owed their positions to the late Leonid Brezhnev, whose policies have subsequently been renounced and even ridiculed in the Soviet press in favor of glasnost and perestroika. In December 1986, Kunaev lost both his First Secretaryship and his position in the CC CPSU Politburo, after which riots had occurred in Kazakhstan, because locals objected to his replacement by an ethnic Russian, Gennadii Kolbin. Shcherbytsky became even more isolated.

Many Sovietologists have maintained—it must be added with little hard evidence—that Shcherbytsky and Gorbachev do not get along. In their view, the Chernobyl accident provided a convenient excuse for Gorbachev to remove finally his 69-year-old "rival."[58] On this same potential "purge" list was found the Kiev Secretary Revenko, who had been in office for only six months when the Chernobyl disaster occurred.

Evidently, either Gorbachev and Shcherbytsky are not at loggerheads, or other factors came into play. Shcherbytsky, in the face of a crumbling Ukrainian party edifice—the Oblast First Party Secretaries in Voroshilovhrad, Lviv, Dnipropetrovsk and Volyn, and the Chairman of the Ukrainian Council of Ministers were all replaced in 1987[59]—kept his position in that year. Some Western journalists have even speculated that Gorbachev feared a Ukrainian backlash *a la Kazakh* if he removed the Ukrainian leader, but this is unlikely. The likely reasons for Shcherbytsky's continuation in his job were both economic and psychological.

While there is little evidence that Ukrainians per se are enamored of their party leader—he is a dull, somewhat colorless character, prone to making longwinded speeches echoing whatever line is in vogue in Moscow at the time, and frequently blaming all his colleagues for Ukraine's economic problems—there is no reason why Moscow should be dissatisfied with the compliant elder statesman in Kiev. Not only did he curb a potentially unruly dissident movement in the 1970s and push the expansion of the Russian language in Ukrainian institutions, he has also supported Moscow's economic initiatives in Ukraine, such as the introduction and rapid expansion of nuclear energy.

Both Shcherbytsky and Secretary of the Central Committee of the Communist Party of Ukraine, Borys Kachura, have been closely identified with the new program, which is one means of ensuring the continued economic importance of the USSR's "second Soviet republic." Had Shcherbytsky fallen casualty to Chernobyl, the public may have perceived his removal as a condemnation of Soviet nuclear power strategy in Ukraine. This is not to say, however, that he was not blamed for the accident. Clearly it appeared that his was the main responsibility, but the political repercussions were deliberately limited to the lower levels of the party.

Following the initial repercussions in the local party organizations, which included dismissal of the party secretary at the Chernobyl plant, Parashin, and the severe punishment of A. Hamanyuk, the First Party Secretary of Prypyat City Party Organization, there were frequent reports about the lack of discipline in the local party organizations. In the Komsomol also, according to one report, many First Secretaries "lost their heads" after the accident, and were unable to take actions appropriate to their status. These people had reportedly shown themselves to be unworthy of the trust placed in them before the Chernobyl accident, but had been "forgiven" and given a second chance.[60] Yet there is evidence that they were "forgiven" after the disaster too.

Thus Erik Pozdyshev, who replaced Bryukhanov as Chernobyl plant Director on May 25, 1986, complained bitterly during an interview in October 1986 that:

We enterprise leaders are perfectly aware of our obligations, but unfortunately we do not have many rights. For instance, we have fired a number of people at the station, they had forfeited our trust and the right to work at the atomic energy station. And now these people are being reinstated, most frequently through the courts. But such workers are not needed at the station, they proved useless at the time of the accident. How can those who fled from the station during the most difficult days be taken on again? Such people have to face the collective at meetings, and the workers ask them point-blank: "Why did you run away?"[61]

According to Oleksandr Kovalenko, the spokesperson for the Kombinat production association, since the accident, 67 Chernobyl plant workers have either been fired or demoted, and 27 were expelled from the party. At least 20 of those dismissed were in the control room of reactor No. 4 when the accident occurred.[62] But as Pozdyshev pointed out, these dismissals were not always of long duration.

The government ministries, rather than the party, bore the brunt of the blame for the accident. The bulk of the firings occurred in the Ministries of Power and Medium Machine Building, which has jurisdiction over those stations used to produce components for the nuclear weapons' program. The jurisdiction of the Ministry of Power and Electrification of the USSR and the USSR State Committee to Ensure the Safe Operation of Nuclear Power Engineering was reduced considerably by the establishment in July 1986 of a new Ministry of Atomic Power

Engineering of the USSR, under Nikolai Lukonin, formerly Director of the Ignalina RBMK-1500 nuclear plant in Lithuania.[63]

By late 1987, it was reported that the Ministry of Power and Electrification of the USSR was still responsible for the construction of nuclear plants, but that control over their operation had been given to the new ministry.[64] In addition to the Ministry of Atomic Power Engineering, an "interdepartmental scientific and technical council on nuclear power industry questions" was established, and subordinated to the USSR State Committee for Science and Technology.[65]

Potentially the most important and certainly the most curious political consequence of the Chernobyl disaster was the trial of plant officials that took place in the town of Chernobyl in July 1987. The location, while symbolic, was also dangerous. At least one of the defendants could not come to trial before the summer of 1987 because he was ill with radiation sickness (Mykola Fomin), and clearly the others could hardly "afford" to take further risks to their health by remaining for over a week in a town that was still in the process of decontamination. Foreign journalists, who had been allowed a considerable amount of leeway in terms of post-accident visits to the reactor site, to Prypyat, and to Chernobyl, were banned from the trial for all but the first and last days.

The Soviet press barely reported on the events other than on those two days, even though it quickly became evident that the real revelations about Chernobyl and the Soviet nuclear industry occurred not at the IAEA meetings in Vienna in August 1986 and September-October 1987, but at the Chernobyl trial. The only Soviet source to report on the trial in any detail was, not surprisingly, *Moscow News*, where Andrei Pralnikov exhibited a healthy disdain for the secretiveness shown elsewhere in Moscow. This signified that the Western public knew more about the Chernobyl trial than the Soviet public, since about 75% of the circulation of *Moscow News* is outside the Soviet Union, and about 50% of the 1 million copies issued weekly are published in English.

However, after the first day, even Soviet journalists were for the most part restricted from entering the courtroom, ostensibly for lack of space. The head of the information department at the Soviet Embassy in Washington, Oleg Benyukh, stated that it was "international practice" to prohibit journalists from certain events, for example, some sessions of the Iran-Contra hearings.[66] Yet Chernobyl, unlike those hearings, did not appear to involve matters of domestic security. Moreover, many journalists felt that the accident was an international rather than a strictly Soviet affair since, as the Soviets themselves had pointed out, "radioactivity does not recognize any national boundaries."

The consequence of the barring of reporters was that even Soviet sources thereafter were obliged to reply on TASS reports. Incidentally, TASS was also the first agency on the scene immediately after the accident. At that time, the agency had provided figures on radiation levels around the nuclear plant that later turned out to be erroneous, as the editors of *Izvestiya* acknowledged.[67] Only when journalists and reporters from newspapers and from the Novosti agency arrived on the scene did a more accurate and detailed picture of Chernobyl emerge. However, the situation was now repeating itself as the world waited for news about the trial.

One Western source referred to the official handling of the trial as "selective glasnost." The article pointed out that in early June 1987, the new Chernobyl Director (after Pozdyshev had "received his rems" and been moved to the Ministry of Atomic Power Engineering in Moscow), Mikhail Umanets, had informed foreign reporters that a large group of them would be permitted to cover the trial which, he felt, would "bring the truth to the whole world." Yet it transpired that only a handful of such journalists were allowed to attend even the first and last days. He noted also the inadequacy of the reports provided by TASS. The main reason for the secrecy, in his view, may have been the embarrassing details that emerged about the RBMK-type reactors.[68]

But was this decision premeditated, or was it a sudden impulse of the authorities? Above all, why was a more appropriate trial site, such as the city of Kiev, not selected instead? Kiev would not have involved dosimetric checks whenever someone entered the courtroom. The original trial had been scheduled for March 1987, thus dosimetric checks would probably have had to have been even more thorough had it taken place then. Holding the trial in Kiev would also have enabled a much larger attendance, but evidently the Soviets were not anxious for a large crowd. Only at a Moscow press conference dedicated to the 30th anniversary of the IAEA on July 27, the penultimate day of the trial, did Radio Moscow announce that 14 foreign journalists were to be permitted to attend.

The defendants were as follows: Viktor Bryukhanov, aged 51, the former Director of Chernobyl nuclear power plant; Mykola Fomin, aged 50, former Chief Engineer; Anatolii Dyatlov, aged 57, Fomin's deputy at the time of the accident; Oleksandr Kovalenko, aged 45, the head of the fourth reactor unit; Boris Rogozhkin, aged 52, the Shift Supervisor on the night of the accident; and Yurii Laushkin, aged 50, the State Inspector at the Chernobyl plant. The most serious charges were directed against Bryukhanov and Fomin. The court was in the charge of Raimond Brize of the USSR Supreme Court, and the State Prosecutor was Yurii Shadrin, Senior Assistant to the General Procurator of the USSR.[69]

Five of the defendants were charged with violating safety procedures, contravening Article 220 of the Ukrainian Criminal Code, which stipulated responsibility for violating rules of technical safety at "explosive-type" enterprises, entailing people's deaths and other severe consequences.[70] The State Inspector, Laushkin, was accused simply of negligence.

Bryukhanov's case was the most explosive and the most disturbing. He had arrived at the plant shortly after the accident occurred. But he had not arranged for radiation levels in Prypyat to be checked because the Chernobyl station did not have the equipment to carry out such an investigation. Sampling measurements of radiation had been taken at the plant, but Bryukhanov "ordered the staff to keep silent" about the results. In the reports that he made out to his superiors, it was stated, the "radiation levels were dozens of times lower."

> Having manifested confusion and cowardice, Viktor Bryukhanov did not take measures to restrict the scope of the accident, he did not activate the plan to protect the population and personnel against radioactive irradiation. With the information he presented, he intentionally diminished the data on radiation levels, which impeded the timely evacuation of people from the danger zone.[71]

The account provided by *Moscow News* was even more explicit:

> Learning that the radiation level considerably exceeded the permissible level, Bryukhanov, having a vested interest—wanting to make everything seem all right in the situation that had taken shape, had hidden the fact on purpose. Abusing his power, he submitted to competent superior organizations data with lower radiation levels. The lack of broad-scale truthful information on the nature of the breakdown led to the contamination of the station personnel *and of the population in the the locality adjacent to the station.*[72] [My italics.]

This account provided the first and thus far only evidence that the nearby population was contaminated after the explosions. The clear implication of the above is that Bryukhanov was responsible for the 36-hour delay in evacuating Prypyat and endangering the lives of its residents by not warning them of the dangers involved. This tells us first of all more about the aftermath of the accident than about Bryukhanov personally. In short, it becomes evident first that the comments made by Leonid Ilyin above, in the light of radiophobia, that there was never any danger to Prypyat residents, have no foundation. At best, the authorities could not have known what the radiation levels were. At worst, they had discovered the true state of affairs and followed Bryukhanov's example by concealing it. There is also some question as to Bryukhanov's sole responsibility in that as firemen and first-aid workers were collapsing with regularity from radiation sickness in the first hours after the accident, one really did not require a dosimetric measurement to realize the dangers involved.

Further, as no radiation measurements were taken in Prypyat, the figures provided at the two major IAEA meetings by the Soviet side—a maximum level of 1 rem per hour in the 30-kilometer zone—could not possibly have been accurate. There were no figures taken at the time of the accident or immediately afterward. The highest levels of radiation therefore were *never* recorded, according to the information supplied about the Chernobyl trial. Because the above facts were incidental to the trial itself, they were rarely emphasized in either the Soviet or Western reports. Yet they were among the most remarkable revelations about the entire Chernobyl affair.

In Vienna in August 1986, the Soviets had also maintained that the smooth operation of the Chernobyl station had induced a state of complacency in many of its personnel and that this lackadaisical attitude had led to the lack of precautions during the fatal experiment. In short, a good past record had caused an attitude of "familiarity breeds contempt." The Chernobyl trial effectively blew aside this argument too by referring to numerous problems and violations of the past.

Thus it was held that "gross violations" of technological discipline had occurred "long before the tragic night of April 25-26." These had simply been hushed up. Several years ago, the USSR State Committee to Ensure the Safe Operation of Atomic Energy had discovered that rules were being broken at the station and recorded them. But the official report prepared by the Committee had maintained that the problems had been eliminated. The personnel, unaware of any of this, "freely played cards or dominoes" during their shifts. The management, with "incredible lightmindedness," and despite the fact that a similar experiment had been conducted there before had produced "a negative result," decided to hold another experiment.[73]

Between 1980 and 1986, according to one witness at the trial, 71 "breakdowns" had occurred at the Chernobyl station, but no research had been undertaken into 26 of these cases. Numerous incidences of equipment failure at the station had not been recorded in the operation logbooks. The fourth reactor had come into operation in December 1983 before the necessary tests on the reactor had been carried out. Tests on the turbogenerator had been undertaken and the results were "not good." However, it was reported, Fomin, Dyatlov and others had given the go-ahead for the No. 4 reactor to be put into a test mode ready for full operation.[74]

On the night of the accident itself, Bryukhanov, Fomin and Dyatlov, and the former head of the reactor shop, Kovalenko, had not coordinated the experiment properly, or taken additional measures to ensure safety. The personnel on shift duty turned out to be unprepared for action in an accident situation. Rogozhkin, upon hearing what had occurred, did not notify personnel at the plant.[75] At 8am on the morning of April 26, workers had arrived for the next shift, and hung around the station "unnecessarily," thereby receiving high doses of radiation.[76] Here is another "unknown" factor. Were the personnel arriving for the next shift included in the number of those at the plant site at the time of the accident in the official Soviet report to the IAEA? The answer is no. But the trial revealed that these workers had turned up for work as though it were a normal working day.

How had such gross violations been permitted? Why was the Chernobyl plant in such a sorry state of affairs by April 1986? The response of Soviet reports has been that the problems arose during the "period of stagnation" in the USSR, the later years of the administration of Leonid Brezhnev.

> Where are the roots of the degradation of thos people [Bryukhanov, Fomin, Dyatlov]—all the way from specialists entrusted with important business down to a serious crime which entailed tragedy? I am sure those roots are in the sphere of morals, in the low level...which was generally typical of productive and public relations existing at that comparatively recent time which today is termed the "stagnation period."[77]

In short, therefore, the entire episode could be explained away as an example of slackness that had occurred during the Brezhnev period, which was now being eliminated by an efficient Gorbachev system, which was being frank and open about the mistakes of the past. Moreover, the entire trial suggested that a few individuals were responsible for the major accident. The sentences they received reflect their alleged responsibility. Bryukhanov, Fomin and Dyatlov received 10-year sentences; Rogozhkin received 5 years' imprisonment; Kovalenko was sentenced to 3 years; and Laushkin to the maximum 2-year sentence for his offense.

Interestingly, the defendants were not prepared to accept their own guilt. Mykola Fomin, who had been ill beforehand, had no wish to appear as a sacrificial lamb, and on the first day of the trial, looking sickly and wretched, he blamed the accident squarely on the faulty design of the reactor. Dyatlov declared that he was not "completely innocent," but concurred that the poor design of the reactor had been the decisive factor. Whereas Bryukhanov, Fomin and Dyatlov had been in custody since the accident, the other three defendants had been participating actively in work around the Chernobyl station, living at Zelenyi Mys, the shift settlement for plant workers located on the Kiev Reservoir.[78]

The London *Daily Telegraph* maintained that the main goal of the trial was to put all the blame on a few individuals, thereby exonerating the country's scientific establishment.[79] Certainly this is what has occurred thus far. Despite Soviet comments that the Chernobyl plant's designers would also be brought to trial, this had not occurred by the end of 1987. The world was provided with six scapegoats, including a Director and Chief Engineer who were not even at the plant when the disaster occurred. Neither Bryukhanov nor Fomin caused the accident, and neither could have stopped it. Also, both were products of a system that had built and operated nuclear power plants in a negligent manner.

Was this system simply a "hangover" from the Brezhnev period? When a similar tripping experiment with "negative results" had been carried out in 1983, Gorbachev's mentor, Yurii Andropov, was General Secretary of the CC CPSU. In other words, the precedent for undertaking an experiment in such circumstances had not even occurred under Leonid Brezhnev, but under his successor.

The political fallout from Chernobyl appears to have achieved one major effect: to divert any culpability from the party hierarchy, in Kiev and especially in Moscow. The Gorbachev administration, it appears, inherited the conditions that brought about this tragedy, while the Soviet scientific elite were hampered by incompetents at the local level. The Lausanne newspaper, *Le Matin*, termed the July 1987 event a "show trial." It maintained that the guilt of the accused had already been decided beforehand and that the punishments handed out were a "matter of form," that is, to satiate Western observers and to reveal to them that after Chernobyl, glasnost had finally triumphed.[80] That there is some truth in the remark is revealed by an account of the trial in a Ukrainian newspaper, which stated that one of its goals was "to prove irrefutably the guilt of the defendants"![81]

But if the July trial was merely for show, then why was the 3-week session held almost entirely in secret? It can be argued that the opening and closing of the sessions were for the consumption of world opinion, but that in the intervening session, a true investigation into the nature of the accident occurred. As stated, even the comments made on the two occasions when some journalists were allowed entry have revealed more about the post-accident situation than the two IAEA meetings in Vienna. It is for the hints and statements about what really happened during the first hours and days after the disaster that the trial has proved useful. Hitherto, many scholars had speculated that the situation might be somewhat worse than the picture drawn by the Soviets in Vienna in August 1986. Now they had proof, while still lacking a clear picture.

Bryukhanov and his associates, it should be stressed, were indeed guilty of major transgressions. But they were not the main actors in the drama, despite the withering depiction of their incompetence in the trial and even in Gubarev's play, *Sarcophagus* (Chapter 4). They had not instigated the crash nuclear power program and they were not in a position to heed some of the warnings about this program that had appeared in the Soviet press. They were essentially technocrats— inefficient technocrats, but technocrats nonetheless.

1. Aerial view of Chernobyl. Six months after the disaster work continues to create total biological protection. October 1986. (SYGMA/Hillelson Agency)

2. Control Room at the Smolensk plant, equipped with an RBMK. 30 April 1986. (SYGMA/Hillelson Agency)

Moscow: The Corridor of Hospital No. 6. (Richard Chamberlain, SYGMA/Hillels Agency)

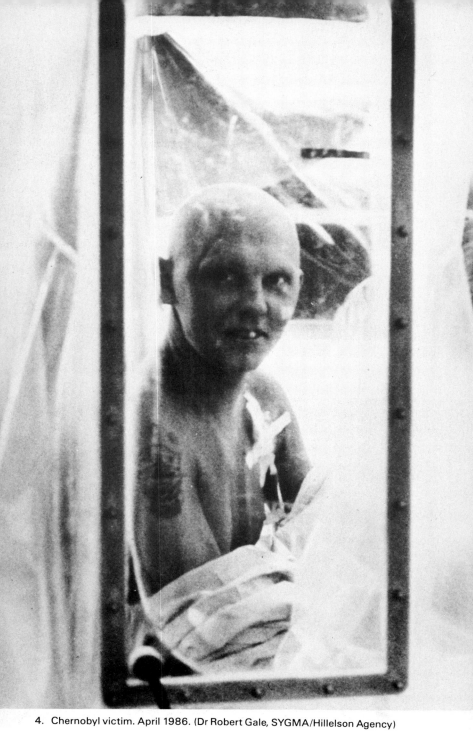

4. Chernobyl victim. April 1986. (Dr Robert Gale, SYGMA/Hillelson Agency)

5. Decontaminating a car. (Steve Rayner, The National Geographic Society)

ДОБРО ПОЖАЛОВАТЬ!

6. The Chernobyl Workers' Rest Centre. (SYGMA/Hillelson Agency)

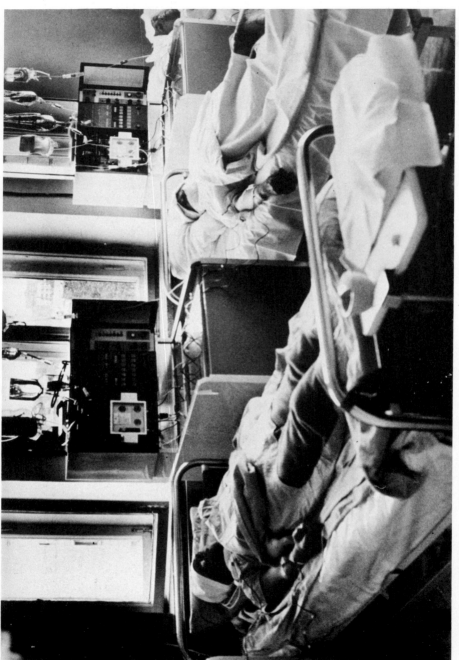

7. Chernobyl victims. (Robert Gale, SYGMA/Hillelson Agency)

8. New houses for Chernobyl. (E. Jacobs, SYGMA/Hillelson Agency)

9. Woman replanting tree at Chernobyl. (Steve Rayner, The National Geographic Society)

4 Images of Chernobyl
ARTS AND THE PUBLIC

Chernobyl was first and foremost a media event. The Soviet authorities have never forgotten the sensationalized accounts that first appeared in the Western press after a nuclear accident had been acknowledged. There were reports from a ham radio operator about dead bodies being piled up in mass graves; the unfortunate United Press International reporter who was informed by a Soviet citizen that there had been 2,000 deaths; accounts of two simultaneous meltdowns based on U.S. satellite information; and banner headlines for more than a week in virtually every Western newspaper. Mikhail Gorbachev's regime had received the sort of attention it had probably craved from the Western media but, as far as the Soviets were concerned, in the worst possible way.

Gradually, the Soviet authorities came to a sort of uneasy compromise with the Western media and with its own public over the Chernobyl disaster. One of the reasons that it is possible to write this book is that the Soviets have ensured that a considerable amount of information has been divulged about the aftermath of the accident. These unusual circumstances for the Soviet Union have been attributed largely to the period of glasnost inaugurated by Mikhail Gorbachev, a leader who is widely respected outside the Soviet Union and who has been concerned to build up his own public image in the Western world.

However, there is more to the question than promoting glasnost.

125

The Soviet Union has indeed opened up its society, often in a very refreshing manner during public discussions, in scientific journals, literature and other spheres. On Chernobyl, a massive amount of literature is available. This chapter will also provide a survey of some of the films and a play written about the event. Yurii Shcherbak has provided us with a wonderful inside view of the lives of the people and some of his interviews will also be cited. But Chernobyl is not officially part of the glasnost campaign, or, to put it more accurately, the openness about certain facets of Chernobyl has occurred without the blessing or guidance of the authorities. It is true that Gubarev's play *Sarcophagus* received official approval and has now been seen in various theaters of the world. But the play echoed official views about the causes of the accident. On Chernobyl, the authorities' steps appear to have been designed, ultimately, to propagate their own interpretation of what occurred at the reactor site and afterward.

This may seem a somewhat harsh attitude. Nuclear power worldwide is a sensitive subject. Would other countries have been any more forthcoming? Were the U.S. nuclear authorities any more open about Three Mile Island than were the Soviets about Chernobyl? Possibly not. But whereas the Western media is and was free to write accounts of Three Mile Island at their leisure, the Soviet press has espoused the opinions of the leadership in Moscow, by and large. Occasionally, a journalist or more likely a group of letter writers have been very outspoken, but subsequently the more "balanced" and reasoned voices of Legasov and Ilyin are interviewed to put matters into perspective.

Additionally, Western journalists have been allowed remarkable access to the Chernobyl plant and its environment. Yet there are sound reasons behind such permissiveness. First, the current Soviet government recognizes well the power of the media in influencing popular opinion. Second, Chernobyl is an international rather than a strictly domestic event. Third, the Soviets have cooperated with organizations such as the IAEA and have been anxious to show that they are working and acting in their nuclear power industry in close cooperation with the experts within that body. It is the first reason with which we are concerned in the first part of this chapter.

Media: East and West

During the summer of 1987, more than 60 "foreign" journalists visited the Chernobyl nuclear power plant. They were permitted to become acquainted with the operation of the first two reactors at the station, to examine the work conditions of their operatives, and to talk with some of the residents evacuated from Chernobyl Raion.[1] The most interesting aspect of the affair was—with all due respect to the Western journalists—how the stories filed by these journalists were received by the Soviet authorities. By the summer of 1987, Soviet sensitivity about Chernobyl, rather than declining, was reaching its peak. Having extended courtesies to the foreigners, they expected appropriate results. For example, the Ukrainian writer Yurii Shcherbak, who himself has provided frank accounts of interviews with personnel from the zone, was furious with an account that had appeared in the *New York Times* under the authorship of Felicity Barringer almost a year earlier:

> I was especially shocked by the cool politicizing tone of the article by Felicity Barringer in the *New York Times*, 5 June 1986: this woman (a woman!) with the sensitivity of a robot, manipulating events with her pen, seemingly slicing up the living with a scalpel, made her report from the pioneer camp "Artek," where at that time the children from Prypyat were located. In her words were no ageold female, maternal charity—only the singularly detestable propagandistic hostility to everything that 11 and 12-year-old children told her, stunned by events, longing for the homes to which they cannot return.[2]

Another reporter who quickly aroused the wrath of his hosts was Mario Dederiks of the West German *Stern* magazine. He had visited Kiev in August 1986 and, according to a Soviet account, had subsequently published a report in which he made reference to obvious exaggerations "drawn from questionable sources" about the contamination of water, silt, and produce, "terrifying the German reader with horrifying prognoses." This was the time, stated the Soviet writer, when there were myths about corpses in the streets of Kiev and thousands of deaths as a result of Chernobyl. However, in the spring of 1987, Dederiks had returned to Kiev and visited the offices of the Kiev newspaper, *Pravda Ukrainy*. We drew his attention, said the writer, to the "disinformational character of these facts." Is he satisfied with the information he is receiving from Soviet experts today? "No, I'm not satisfied," responded the unabashed Dederiks, "there is not enough new information."[3]

Dederiks also featured in a report about what was regarded as a scandalous photograph in the Swedish *Goteborg Post*. According to a report in the Ukrainian press, the Swedish journalists had arrived at Zdvyzhivka, a village for Chernobyl evacuees. They had reportedly photographed one of the villagers. "He was wearing a good suit." But later when the Swedes had sent over a copy of the newspaper in question, what appeared was "not a photograph, but a caricature, "he looks like he's in rags." The interviewee, Anatolii Chernenko, produced the copy in question. On the front page, continued the Ukrainian account, was a large photograph of Zdvyzhivka. But it had been taken of the backyards, with latrines prominent.[4]

Dederiks was also shown the photograph and asked how he evaluated the work of his Swedish colleagues:

> Dederiks: I don't see anything wrong with it. The point is that a panorama of the population point should not have been photographed any other way.
> Chernenko: But our people were offended. They were insulted by such a reproduction of their village.

The photographer for *Pravda Ukrainy*, P. Prykhodko, had also photographed Zdvyzhivka. He had not avoided the backyards, stated the report, but nonetheless his pictures had not turned out badly. One had been handed to Dederiks, but the German had said that it was "inappropriate" for *Stern* magazine. Why?, the Soviet reporter wanted to know. Was it only appropriate to show the worst side of everything?

The criticisms of Dederiks and the Swedes were mild compared with those of the Munich-based radio stations, Radio Free Europe and Radio Liberty, which were partly responsible for the first news about the disaster reaching many Soviet citizens. A former employee of Radio Free Europe who had returned to the Soviet Union was featured in the Lithuanian press, and he accused some of the employees of being disappointed that there were so few victims of Chernobyl:

> At RFE [Radio Free Europe] the professionals of American propaganda worked—unconcerned about the affairs of Lithuania or those of other nations. Even plain objectivity did not worry them. For example, at the time of the accident, a considerable number of workers at RFE and RL [Radio Liberty] even demonstrated happiness at the occurrence, regretting for a long time that there were so few victims. Thus without reservation, they made use of every possible rumor and invention, and the number of victims swelled to 15-20,000.[5]

It is perhaps to be expected that the USSR's surrogate radio stations would be treated harshly in the Soviet press. Nevertheless, the criticism, at least from the perspective of the materials published by RFE/RL, was unwarranted. In fact, RFE/RL seems to have been at pains to avoid the sensationalized accounts that appeared in a few Western newspapers. Yet the quotation and the disputes with Dederiks reveal that the Soviets became very unhappy about the Western interpretation of what had been seen at the beneficence of the Soviets.

What was objected to the most were impressions that the Soviets felt were misleading, such as Prypyat as a "dead city," rags floating on balconies that were the remnants of washing hung out by citizens just before they had been given the order to evacuate. These sort of images were referred to as a "disinformational videosequences."[6] But there were also visitors to the Chernobyl region from the West who reacted in the opposite fashion, lauding Soviet reactions to the major accident, and praising the Soviet authorities effusively. A correspondent for the *Chicago Tribune* was quoted on Radio Kiev in July 1987 as stating that:

> In my opinion, the majority of people in the West, as well as many Soviet people, were distressed by the silence of official circles in the first days after the accident. But since the decision was made to apply glasnost to Chernobyl, the Soviet press has done absolutely fantastic things. The Soviet press has almost deprived we Western journalists of our work, because it illuminated the events so well that the best we could do was to copy the Soviet press every day and tell it to our readers. *Pravda* said everything, *Izvestiya* said everything, and *Komsomolskaya pravda* said everything.[7]

The above is an extreme example. One can hardly imagine that the majority of Western correspondents were reduced to copying official Soviet accounts if they had the opportunity to visit the area themselves. But there was certainly a tendency to praise Soviet reporting in view of the initial silence about the accident in the first days. Fred Jarvis, the Chairman of the General Council of Trade Unions of Great Britain, who headed a delegation to the Chernobyl nuclear plant in April 1987, declared that in terms of safety precautions "we are significantly inferior to our Soviet colleagues" because many of the prophylactic measures at British nuclear power plants are of a temporary character.[8] Such remarks were always repeated in Soviet accounts about the visits of foreigners. They were appreciated, just as descriptions of rags flying on empty balconies were resented.

However, Soviet openness on the issue which, as has been suggested, was directed toward a certain interpretation of the outcome of Chernobyl, evidently changed in the spring of 1987. It became noticeable in the Ukrainian press that reports from the plant site were being written not by journalists, but by the military. Some of the writing was verbose and awkward after the flowing style of the summer of 1987. Shcherbak and Honchar had been replaced by colonels and lieutenants. Harold Denton, Director of the Office of Reactor Safety at the U.S. Nuclear Regulatory Commission, described how during his March 1987 visit to the first two Chernobyl reactors, he saw soldiers guarding the plant everywhere he went.[9]

The importance of the military in the zone had been evident for some time. But increasingly, journalists were being kept out of the 30-kilometer zone. In late April, an article in *Moscow News* complained that it was becoming harder to obtain information from Chernobyl. "The formula 'not for the press' is being used more and more often." The question was raised why information that was being reported on a regular basis to the IAEA could not be used by Soviet journalists. The rumors that develop about such matters, the article continued, were only exacerbated and not eliminated by the absence of summaries and figures.[10] Asked about this report, Chernobyl's Director, Mikhail Umanets, replied that the criticism was unfounded. But, in his view, the average of two delegations that visited the station daily was an impediment to work.[11] His statement was a nonsequitur, however. *Moscow News* had not commented on delegations to the plant, but only about the information question. The latter could have been ascertained by a telephone call had the authorities at the station been willing to discuss many of the issues.

During his visit to the area, William Eaton of the *Los Angeles Times*, questioned by a Soviet correspondent, also stated that it was unfortunate that Western journalists had received so little information about Chernobyl. He was instantly referred to the example of Three Mile Island where, even though the accident had occurred several years ago, neither specialists from Chernobyl nor Soviet journalists had been permitted to visit, whereas Eaton had visited Chernobyl, the nuclear plant and Prypyat.[12] Was the comparison a fair one? Was Three Mile Island an international incident? Perhaps of more concern was that the initial openness about the accident, whatever its limitations, appeared to have dissipated. It was as though a decision had been made: enough is enough. It is understandable that Umanets and company were weary of visitors, but of more concern was the fact that the flow of information from the area had very much dried up.

Visual Images

The first literary output to emerge from the Chernobyl accident was a play by the science correspondent of *Pravda*, Vladimir Gubarev, named *Sarcophagus* after the concrete shell that was used to seal the fourth reactor. It appeared in excerpts in *Sovetskaya kultura* as early as September 1986, and caused something of a sensation upon its appearance by its heavily critical nature. The setting for the play was a radiation clinic, and the main character—the one who acts as the mouthpiece for the author—is known as Immortal, who has received a heavy dose of radiation from an earlier incident in a nuclear laboratory, yet somehow has remained alive for 487 days after exposure. Immortal is the only one to remain alive throughout the play.

During the course of conversations, with characters such as the plant Director (the plant in question is obviously Chernobyl, although it is not given this name), a geiger counter operator, a general, an American doctor that may have been based on Robert Gale, a procurator and others, Gubarev manages to criticize severely the state of affairs of the Chernobyl station at the time of the accident. He refers to an earlier accident regarding the flammable roofing material that was subsequently used for the reactor building; he notes how the plant Director made sure that his own family was safe before he attended to the needs of fellow workers. During the long discussions about responsibility for the accident, Gubarev implies that the decision to dismantle safety devices, such as the emergency cooling system, was responsible for the conflagration.

The key question is what Gubarev intended to do with the play. It was written in seven days and before the information divulged to the IAEA in Vienna (let alone that at the Chernobyl trial) was brought to light. It could not therefore stand as an account of what happened during and after the accident, but only as a writer's first impressions of what had occurred, based on conversations in the area. The accident, strictly speaking, had not arisen because the safety devices had been switched off, as the play implied. These actions made the original experiment less safe, but they did not cause the disaster. Gubarev does, however, provide a biting portrayal of a physicist who believes in the infallibility of the machine.

Moreover, the fact that the play was shown not only in the Soviet Union—it premiered in Tambov in November 1986—but also in Western Europe in the spring of 1987, suggests that its message was one that the Soviet authorities were not reluctant to disseminate.

Ultimately, that message was not about the lamentable state of affairs at Chernobyl, but about the horrors of nuclear war. The 48-year-old author was quoted by one source as stating that "I wanted to make clear to people that they are living in a nuclear era."[13] It could be argued that by including the powerful underlying message about nuclear weapons that Gubarev had succeeded in spreading the message about Chernobyl, but it is difficult to avoid the impression that there is a strong ideological content to *Sarcophagus*.

Further, the secondary message contained in the play is not so much about affairs in 1986-87, but rather about inertia and bureaucratic bungling at the station in past years. It can be interpreted as an attack on the Brezhnev period, when the various abuses came into being. On a literary level, by which all plays must be judged ultimately, *Sarcophagus* probably failed. It was more a documentary than a work of art. Jon Peter of the London *Sunday Times*, referred to the play's "primitive ideolological graffiti" and felt that the play did not criticize or even question the basis of the Soviet system.[14] The play's language is that of an engineer rather than one endowed with notable literary talents. And yet *Sarcophagus* was a significant departure from the norm. In short, anti-nuclear propaganda aside, it was a critical account of the situation at Chernobyl before the accident. It went further than L. Kovalevska, the Prypyat journalist who had been demoted for writing the criticial article about the situation in *Literaturna Ukraina*. It can be argued that as a critique, it still did not go far enough. Yet it went much further than previous critiques and Gubarev deserves the accolade of a pioneer in this respect.

Of the films to appear about Chernobyl, one of the first (most were delayed for various reasons, about which more below) to appear was *Kolokol Chernobylya* (The Chernobyl Warning Bell), by V. Sinelnikov and R. Sergienko, and the cameramen were Konstantin Durnov, Volodymyr Frolenko and Ivan Dvoinikov. It was produced by the well known Soviet producer, Dzhemma Firsova, a laureate of the Lenin Prize and holder of the State Prize of the USSR. In contrast to Gubarev's play, *Kolokol Chernobylya* was hardly an expose and dealt little with what had actually occurred at the station to bring about the explosions. It began with Valerii Khodemchuk, one of the operators at the station believed to have been killed instantly at the time of the accident:

Well he went out into the street and then sort of looked around the yard: "So, do svidannya, mother." I said "Have a good trip." And then he simply drove off. And I never saw him again, my little son.[15]

As the excerpt reveals, the film focuses very much on the human side of Chernobyl. It was made between May 28 and June 26, 1986, and some parts were added in September 1986. Many of those featured in the film participated first-hand in the events. One who did not was Western industrialist Armand Hammer, who also appears in the film to comment that when Mr. Gorbachev meets with Mr. Reagan, he will tell him all about this film.

The emotional nature of the film is evident throughout and it evokes pathos with the simple actions of uninformed peasants returning to their contaminated farms or the arguments between officials at the plant. Others were fishing in a contaminated lake, and informed the interviewer that their fathers and grandfathers had fished here, so they were going to do the same. Servicemen were given the following briefing before they began work:

> When you set off from here, begin counting to ninety, then drop your barrows and spades and run back as quickly as possible.

Were emotions a sign of weakness in the Soviet state? No, responds one critic, because it was Lenin who said that "without human emotions there never was and never can be a human search for truth." Behind the "new thinking" initiated under Mikhail Gorbachev lie these emotions: guilt, burning shame, repentance; in complacency lies evil:

> Chekhov has the tendency to hang above the door of every lucky man a hammer, to remind one of the misfortunes of others. Here, it is not a hammer, but a bell. No one asks for whom it tolls. And its toll is not funereal and not deafly hopeless, but forestalling, calling people together....It is the thinking of the future, without which we cannot make one step into tomorrow.[16]

According to one critic, the merit of *Kolokol Chernobylya* is that it depicted the simple people, those whom scientific progress had passed by, but at the same time who suffered "from our incompetency in scientific-technical progress." He felt that the film revealed the horrors of the accident in bringing into reality the science fiction of the past. Just as the fantasies of Jules Verne and H.G. Wells had been realized "almost 100%," this film showed that the writings of Ray Bradbury were now being enacted in reality, "the inevitable has occurred!" He cited the following passage, which is here translated from his original Russian into English rather than directly from Bradbury himself, since it is obviously the translated version that had significance for the critic:

> People who live here will remember their native homes as the distant and irretrievable past.
> And no one will gather these apples. They will rot, retaining inside them their radioactive seeds. Wheat kernels will spill to the earth, and grow, and again a radioactive field will sprout. Birds will peck at them—you cannot establish boundaries for them.[17]

The film, then, brought home the human side of the accident. Evidently it took four months before the State Film Board cleared it for showing, and it appeared at the West Berlin film festival in early March 1987, where it reportedly had a mixed reception as a result of the relatively poor quality of the filming (there were radiation spots on the film itself, which some Soviet sources felt only added to its drama).[18] One reason for its delay may have been the emotions it obviously aroused in those who saw it. For example, it prompted E. Alekseev of *Literaturnaya gazeta* to criticize bitterly the Soviet Academy of Sciences, from the viewpoint of the artist who cannot comprehend the apparent callousness of scientists:

> 31 perished, 28 from the radiation blast. Is that many? Few? What is the arithmetic about? What is this blasphemous "science demands sacrifices," the dictum of the callous soothsayer of the Academy? We don't need measures, the supply of inadequate equipment to the atomic energy stations and projects that risk people's lives for the sake of economic ends, and operators capable of making 5-7 errors. And leaders who experiment in a situation where only the strictest letter of the law is to be followed.[19]

Implicit in this critique is the author's bitterness regarding the frequent reiteration of the fact that the accident had caused "only" 31 deaths (regardless of the accuracy or inaccuracy of this figure). *Kolokol Chernobylya* had shown that these people were helpless victims and that they had died unnecessarily through the carelessness and even heartlessness of others.

But there were other reasons for the attempted suppression of the film. According to V.G. Afanasyev, the Chairman of the Board of the Union of Journalists of the USSR, the film had transgressed certain "forbidden zones," in which criticism was still not permitted. It had been "demanded" that all words such as "tragedy," "misfortune," "catastrophe," be removed from the film. In Afanasyev's view, the lesson to be learned from such censorship was "the lesson of lies." It was not as though removing such words would change the essence of the film. He himself had noticed that those portrayed cleaning up after the accident were improperly dressed for the task. This was unfortunate for those in question, but in his opinion should not have resulted in a coverup. Rather the departments concerned with the distribution of protective clothing should have been admonished.[20] In other words, the Soviet authorities did not wish to disseminate a film that provided evidence that cleanup workers did not have proper protective clothing.

Another film that made its debut in March 1987, in the raion center of Brovary, Kiev Oblast, was *Chernobyl: khronika trudovykh nedel* (Chernobyl: A Chronicle of Difficult Weeks). This was a Ukrainian film made at the Ukrkinokhronika studios in Kiev. Its producer and camera-operator was Ukrainian film director Volodymyr Shevchenko, accompanied for the most part by two cameramen, Viktor Krypchenko and Volodymyr Taranchenko. According to Radio Moscow, it was a film about the builders and firemen, the doctors and the cooks, the coal miners and truck drivers, helicopter pilots and bulldozer operators, and about those who accepted evacuees into their homes.[21] But the film, whatever its merits and demerits, will retain an unending significance as a result of one fact: Director Shevchenko died of radiation sickness after it was completed and in fact did not live long enough to witness its premiere. Shevchenko was the thirty-second casualty of Chernobyl, according to a Soviet source.

Shevchenko's crew was the first on the scene after the Chernobyl accident. Initially they were refused permission to film in the area, but early in May, this decision was rescinded and the crew, with brief respites in Kiev, actually lived in the contaminated city of Chernobyl from May to August 1986. "In those days," stated one source, these artists had put aside all their other work. Their one purpose was to tell their compatriots "the truth."[22] To find out more about the personalities involved in the making of the film, a Ukrainian newspaper correspondent interviewed the film's narrator, I.Yu. Malishevsky.[23]

Malishevsky had first met Shevchenko at the nuclear plant itself. During the early days of the filming, the goal was not to omit anything in the area of the damaged reactor. He stated that at the time of the filming Shevchenko was already seriously ill—a statement that today coincides with the official line about Shevchenko's death, but which contradicts other, earlier statements about the causes behind his sickness.[24] Shevchenko "worked selflessly," he did not spare himself. By October 2, the film was ready.

Malishevsky was asked about his collaboration with Shevchenko on 10 different films, notably three films on the Second World War. He replied that Shevchenko was "a great man," that he loved his trade and that many were attracted by his enthusiasm, honesty and unselfishness. Taranchenko also had a long experience with the Ukrainian film studio, dating back to 1958, while Krypchenko had been working there for twelve years. In brief, a highly experienced crew was in place and through their bravery and resilience, 20,000 meters of film had been shot and were ready for showing a little more than five months after the accident.

At this time, Shevchenko's problems began. Like *Kolokol Chernobylya*, *Khronika trudnykh nedel* ran into difficulties with the authorities. The film was cut drastically, controversial parts were simply edited out and, in the words of one Soviet writer, "the habit of passing off what one wishes for the truth proved to be extraordinarily strong."[25] Of the 20,000 meters, only 1,500 meters went into the film that was eventually shown six months later. The traces of radiation that had appeared on the film, and evidently created a sensation when they appeared, were "simply scrapped." Shevchenko was obliged to rework his film four times to remove various scenes. Unfortunately, stated one source, those responsible for delaying the film's completion were not seen on the film (members of Goskino, the USSR State Committee for Cinematography).[26] Dzhemma Firsova attended the film's press screening and asked whether the film *Razgrom nemtsev pod Moskvoi* (The Defeat of the Germans at Moscow) would have been as effective had it appeared six months after the decisive events. Yet it was actually almost a year after Chernobyl before audiences got a chance to see Shevchenko's work.

In March 1987, Shevchenko died. He did not live to see the final version of his work. Despite his efforts, his film had been dealt with roughly, heavily edited and delayed. "A certain stagnation" had not yet been eliminated by Gorbachev's policies of "glasnost" and "perestroika." Instead, more than 92% of the footage shot by the Ukrainian film crew has not been shown to Soviet audiences. Yet it contains possibly the most important scenes of the events that took place immediately after the accident. Whether these scenes will eventually be made into a second film, as the film-makers have suggested, or whether they will be consigned to the archives indefinitely remains a moot point.

Another film that ran into difficulties with the authorities because of its frankness and exposure of some of the difficulties involved in cleanup work was *Chernobyl: Dva tsvieta vremeni* (Chernobyl: Two Colors of Time), made by the Ukrainian telefilm association, and eventually shown on Moscow television on the first anniversary of the disaster, having been seen earlier in Kiev. It was authored by L. Muzhuk and Kh. Salhanyk and directed by I. Kobryn, with the main camera work being carried out by Yu. Bordakov approximately one month after the accident.

It was said that the film not only reflected "new thinking" in the Soviet Union, but was also experimental in nature. A complete shift of cleanup work was filmed, for example (which would only have lasted a few minutes), which showed the futility of removing radioactive debris from the roof of the reactor with shovels. Those parts of the film that have appeared in the West (there were three parts in all) are set to eerie music and the silhouetted cleanup workers are shown flinging away the debris and running from the scene as fast as their protective clothing permitted (in this case they at least had the proper clothing).

The film "rejected hierarchy." It included "in one sweep" the assembly worker and the Chairman of the Government Commission, the scientist and the construction worker. During the post-accident situation itself, according to one account, bureaucratic problems had emerged and had had to be overcome, something that the film showed honestly. Nevertheless, it did not have a smooth passage onto Soviet television:

> *Chornobyl: dva koliory chasu* is a film that...means so much for our television. And for us, the viewers....It is a chronicle and a warning, a professional work of the highest class by those filmed and those who did the filming....One begins to grieve along with the authors that this exceptionally important, principal, authentic work came to the television screen only several months after its completion.[27]

Other Soviet documentaries and films have also appeared, most notably perhaps *Preduprezhdenie* (The Warning), an 85-minute documentary that was the first to appear on Soviet television, in mid-February 1987. *Chernobyl. Osen 1986* (Chernobyl. Autumn 1986) examined the entombment of the Soviet reactor. The State Committee for Cinematography had prevented its release on the grounds that it manifested "excessive optimism." As *Pravda* pointed out, this was a strange reason for prohibiting a film because it suggested that only critical films were worthy of attention.[28] Goskino had also stated that the film was too traditional and that similar documentaries had already been made. The film, like virtually all the others, had been screened only after the greatest of difficulties.

Nevertheless, taken together, the films and documentaries captured the essence of the situation at Chernobyl in the first weeks and months after the tragedy. The filmmakers had to work in dangerous, often appalling conditions with no certainty that their efforts would be rewarded by having an unedited version appear for the sake of posterity. The scenes were compared in their depiction of heroism to those about World War II, and it is because of this comparison that they were eventually shown in the theaters and on Soviet television. They have clearly evoked criticism in some circles of both nuclear power and of those who maintain that science demands some sacrifices. Had the accident occurred five years earlier, it seems safe to say that such films would never have been shown either inside or outside the Soviet Union. Even in the era of glasnost, when the filmmakers have powerful supporters, the majority of the scenes filmed will never been seen by Soviet audiences.

The Evacuees

The Chernobyl disaster dramatically changed the lives of thousands of residents of northern Ukraine and the southern part of Belorussia. Without prior warning, these people were suddenly ordered to pack their valuables and to prepare for an evacuation on the afternoon of April 27, 1986. They included nuclear plant operatives, farmers, women and children, and most notably the elderly who were the most reluctant to leave homes that in many cases they had occupied all their lives. The official Soviet account of the evacuation was that it was a smooth process, that the evacuees were taken care of financially and materially by the Soviet state, and were given a warm welcome in their new, temporary homes. The unfortunate fact is that the evacuation could not have been smooth under any circumstances. The evacuees were more profoundly affected by the accident than any other group of people with the possible exception of the cleanup crews. Chernobyl changed their lives and only by looking at some examples among them can one ascertain an accurate picture of the social impact of Chernobyl.

The accounts that follow stem exclusively from Soviet sources. Some were described by the Ukrainian writer Yurii Shcherbak in the summer 1987 issues of *Yunost*, but the majority were simply statements and brief accounts that appeared in the Soviet press at various times. These accounts, by and large, were not particularly objective. The reporters and correspondents never attempted to conceal their sympathy for the evacuees and generally tried to convey the impression that they were received with the utmost warmth and hospitality. This author tends to believe these statements, objective or not. Ukrainian peasant hospitality is well known even in the West. My aim, simply, is to provide a general impression of what happened to the people and how they felt about their new situations, within the general spectrum of the aftermath of the disaster.

Initially, the evacuees arrived on buses from the villages within the Chernobyl reactor zone. Some arrived at the "Gorky" collective farm in Makariv Raion, where Mykola Tyapko, Collective Farm Chairman and Secretary of the farm's party organization had spent a sleepless night making arrangements for their rehousing. Two Chernobyl villages, Chapayevka and Horodchany were being resettled in the villages Fasova and Lyudvynivka, and residents of the latter—"it is no exaggeration"—greeted the new evacuees like family. Two deputies from the local raion Soviet committee went onto each bus to direct the drivers. As the bus approached a house, the deputy on board would shout, for example, "There are three places in this home. Who wants to stay here?"[29]

Tyapko himself acted as host to the Chairman of the "Chervonyi Polissya" collective farm, Aleksei Kirilchuk, with his family of five, who had to be accommodated in a single room. The process of allocating rooms in houses took about 2 hours and 30 minutes. In some cases, local residents actually gave up their entire homes to the evacuees. One example is cited in which a farm machine operator and his wife, who had just built themselves a new home, gave it up to a family of eleven and went to live for the time being in a storehouse.

But not everyone received housing so quickly, albeit with great inconveniences. In March 1987, the following letter appeared in the newspaper *Izvestiya*:

From the Tereshchenko family, Stavropol:
In its misfortune, the Tereshchenko family turns to you. We are former residents of Pripyat. After the accident at the Chernobyl nuclear power plant, we received an assignment in the city of Stavropol from the USSR Ministry of Power. We turned to the City Council about the distribution of housing—a refusal.

Mr. Tereshchenko went to the Council of Ministers of the

Ukrainian SSR where he was provided with a letter authorizing the receipt of an apartment. Subsequently, after it had been received by the Stavropol district executive committee, we learned that our problem had to go again before the City Council. At the meeting with the Chairman of the City Council, N.A. Maslov, they refused to enlist us. "You don't live in our city!" Yes, our residence permit is from Pripyat. But you see such derision for people who have fallen into misfortune!

We correspond with our former neighbors, friends from Pripyat. They have all received housing already. But we cannot even believe anymore that we will eventually have a shelter over our heads.

A genuine one, I mean, and not just promises of one.[30]

The Tereshchenko case suggests that those transferred for work in other cities may have found their situations more difficult than those who moved simply from one farm to another. One case, which would have been amusing under other circumstances, was cited by Shcherbak. The Miroshnichenko family of four was moved to the city of Yalta in the Crimea, in accordance with directives of the Ministry of Health Protection of the USSR, issued on September 6, 1986. Upon their arrival in Yalta, the local City Council sent a letter to V.I. Voloshko, the Chairman of the Prypyat City Soviet. The letter requested that the Prypyat Council send to the Yalta Council a certificate confirming the fact that Miroshnichenko had given up his apartment in Prypyat! Evidently, it was felt in Yalta that Miroshnichenko was evading normal rules for the receipt of an apartment.[31]

Another example of post-evacuation problems occurred in the Rovenky area of Ukraine (Voroshilovhrad Oblast). Here, a wife whose husband was still hospitalized as a result of the disaster found herself alone in a fairly remote village with two children, one at school and the other a baby. It was proving impossible for the woman, T. Lyamina, to obtain milk for the baby. The village was a long way from the raion center where the milk could be purchased and the buses ran irregularly. When the other daughter was at school, there was no one with whom the mother could leave the baby. However, the Chairman of the Rovenky City Party Committee, V. Ponomaryov, had evidently misinterpreted the woman's predicament and had claimed that there could be no question of advantages for the children from Prypyat. The matter had subsequently been referred to the Oblast Executive Committee in Kiev.[32]

Even relatively prominent people had their share of problems.

Shcherbak has described the plight of Lyubov Kovalevska, former reporter for the newspaper *Trybuna energetika* in Prypyat, and the author of the best known critique of the Chernobyl nuclear plant before the accident, and now employed with the magazine *Raduha*. Having been evacuated initially, Kovalevska left the new village in Poliske Raion for Kiev on May 8, 1986, having sent her mother and children to Tyumen, in western Siberia. She had distributed her remaining money to Prypyat residents at the airport out of sympathy for the weeping mothers and children who were being moved out of the republic. From the airport at Borispol to Kiev cost 80 kopecks from her remaining ruble. Thus she had 20 kopecks left and described herself as "dishevelled" and "confused."

Having called various friends from a taxistand in Kiev to find that they were not at home, she decided to take a cab and to tell the driver that the friend at the destination would cover the fare. While she was making this decision she was approached by a man who asked her the time. During the conversation that ensued, he established that she was from Chernobyl and destitute and took her by the hand. She misunderstood his motives, "this man is going to take me to his place and so forth." However, Kovalevska was fortunate. The man paid for her taxi and for her hotel room, and thus her problems were resolved.[33] One can imagine from the above some of the difficulties that must have been experienced by those who went to find new places of employment and, unlike the farmers, could not simply rely on state help for their needs.

Old people did not want to move at all for the most part. An instructor at the Prypyat City Committee Komsomol Organization remarked that pensioners were often not evacuated for days, "I took one pensioner out on 20 May." In one case, a man who had taken part in the battle of Stalingrad visited army headquarters, took a respirator and some other items and stayed in his home, not turning on his lights at night. He was eventually found because his son alerted the authorities after the city's water supply had been turned off. These people were known locally as "partisans."[34]

Writing in the Estonian Komsomol newspaper, *Noorte haal*, in August 1986, the journalist Tonis Avikson also described how old people stubbornly refused to be evacuated:

During the evacuation, many old women and men acquired a different wind. They fled to their home forest, concealed themselves in cellar nooks, lofts, cucumber salt water barrels, and even under firewood.

Even when they succeeded in fishing them out from there and taking them away, a day later, or 2-3 days later, they were back again, sitting like a pair of cuckoos on a bench in front of their cottage. Under the concealment of night, they knew how to slip past the control posts. In this way, the process was repeated on several occasions. The old women and old men implored the soldiers in God's name: Do not take us away, sons. We are already old, and have seen everything in life. Even the Germans were unable to take us....Let us at least die at home. Wherever it was in the least possible, they got their wish.[35]

In the early spring of 1987, frustrated by the lack of news about their eventual return to their homes, two grandmothers from the village of Ladyzhychi, about 25 kilometers (15 miles) from Chernobyl, who were aged 70 and 74, decided to take matters into their own hands. Having suffered through an entire winter away from their homes, they went to the raion center Ivankiv fairly close to their temporary home. From there, they took a bus southward to Kiev, a journey of more than an hour. From Kiev, they headed northward again, by train, to Chernihiv, which is located about 170 kilometers (110 miles) to the northeast of the Chernobyl nuclear power plant. From Chernihiv, they managed to get rides in cars and horse-drawn wagons to the Dnieper River, close to the village of Teremtsi, in Chernobyl Raion, and took a ferry across the river.

However, in order to get to Ladyzhychi on the peninsula, it was necessary to avoid all the military controls, "because, you know, we had no permits, no consent." Finally, the two old ladies arrived back in their native villages:

"And what did you do in Ladyzhychi?"
"We whitewashed our houses, put our homes in order, and made them look good. And while we were cleaning, we sang, we sang songs of spring to our God."[36]

The incident encapsulates the feelings of many elderly people, although not all were able to emulate the example of the two Ukrainian grandmothers. Some were interviewed by Western correspondents who had been permitted to visit them in their new homes. A 73-year-old woman in Zdvyzhivka, Anastasya Remizenko, living some 70 kilometers from the damaged reactor declared that she was old:

I miss the old wood stove where I used to curl up at night. These [new] houses don't have them. They are warm, but it's not the same thing.[37]

In contrast to the predicaments of old people, parents were understandably anxious to remove children from the 30-kilometer zone as quickly as possible. One account came from Aneliya Perkovska, the Secretary of the Prypyat City Komsomol Committee, who had the unenviable task of deciding which children were eligible to go to pioneer camps, and whether they would be going to the Artek camp or the less popular "Moloda hvardiya" camp. The main problem was that only children between the ages of 7 and 15 were allowed to go. To circumvent the age restrictions, Perkovska doctored some of the children's birth certificates:

> The arrangement was as follows: to take to camps those who had finished Grade 2 up to Grade 9, inclusive. And they [the parents] came up to me and said: "And the tenth graders—are they not children? And what do we do with the first grade?" So imagine: a mother arrives, and her child is 6. And what is he supposed to do, finish the second grade without fail? What is she going to do with him? Naturally, I go ahead and write, without a twinge of conscience, a different year of birth for this boy. Later when I went to pioneer camps, I sensed a lot of reproach directed at me. But forgive me, I had no other solution.[38]

Later it became apparent to Perkovska that the transgressions from other regions were plentiful. Kiev residents, for example, anxious about the safety of their children, telephoned Prypyat to try to get their children's names on the official lists. Consequently, the Prypyat Committee announced over the radio that parents had to appear personally with their passports and present their Prypyat residence permits. When Perkovska visited the "Artek" and "Molodaya hvardiya" camps in August 1986, she discovered a girl from Gomel there and another from Poltava Oblast. When danger threatened, all the children were suddenly "Prypyatites"!

Some of the children from the Belorussian side of the 30-kilometer zone were sent as far away as Latvia, a daunting experience for youngsters, some of whom had not left their homes before. That their reception was a warm one is evident from a letter from a Belorussian parent that appeared in *Izvestiya*:

> Not only my family, all our Malavtyukovskaya school with the same words of praise recall today Sarma Keish, the Director of the pioneer camp at the Latvian city of Yurmal. With such true motherly love

were met there our children, who had never before been so far from their families. I thank everyone who were toward them—children from Belorussian villages—kind, sensitive, attentive. When one meets such true communists as Sarma Yanovna, one starts to say that all people are good, that generally evil does not exist in the world....A low bow to all those who helped our children.[39]

Although on the Belorussian side of the border, the removal of children from the zone took place later than in Ukraine, it soon approached an enormous scale. According to the Chairman of the Council of Trade Unions of Belorussia, V. Honcharov, 57 sanatoriums, dispensaries, recreation centers, 71 labor and recreation centers, and 32 kindergartens in various areas of Belorussia were "immediately" made available to evacuated children and to pregnant women. Staff were set up in each oblast of the Belorussian republic to deal with the needs of children, and up to 10,000 children were sent daily to the recreation centers by special trains.[40] Nevertheless, on both sides of the border there were reportedly "endless lines of people" who during the evacuation process had lost their children, their relatives and their friends.[41]

At times, mothers even in areas that were not, strictly speaking, in the danger zone, were in a state of panic in their anxiety to get their children out of the area. For example, a 30-year-old Kiev teacher was interviewed by a Ukrainian newspaper, and asked whether her children left Kiev immediately after the accident—a time incidentally when the Soviets had reported there was no danger of radioactive fallout to Kiev citizens:

I have a 10-year-old son Oleh, a 5-year-old daughter, Halyna. After the accident there were many rumors as to the probable evacuation of the population. I felt strong anxiety for the safety of my children. Frankly speaking, I was in despair. I phoned my acquaintances in Moscow. They immediately agreed to take Halyna in. However, it so happened that neither my husband nor I could leave work or go on short leave. That time our friends were of help again. Our neighbors went to Moscow and they took the girl with them. Then, I worried about the son. I calmed down only after all the students of the school had been sent to Young Pioneer Camps in Poltava Region.[42]

By September 1986, the children had returned from the camps and were living with their parents in the new locations. The most immediate problem was providing sufficient teachers and schoolbooks for them, while the schools themselves were somewhat crowded. Some of the Prypyat teachers had fled the area after the accident, joining the mass exodus of those who did not wish to wait for an evacuation to be announced. As a result, students who had completed five courses at pedagogical institutes of the republic replaced them. Tenth graders from the middle school No. 4 in Prypyat no longer had textbooks, but had to write their final examinations. Therefore the school's staff proposed that they set their own grades. Reportedly not one pupil had raised his or her grade in the process, and in several instances, the children had lowered their grades below their normal average.[43]

For pregnant women (described in Chapter 1), their reception at health centers often resembled that of Biblical lepers. In May 1986, for example, some pregnant evacuees were driven to Belaya Tserkov health clinic and the staff came out to meet them wearing gas masks and protective clothing, and checked the women for radiation levels on the street before they even entered the clinic. The same applied to those who were hospitalized after the disaster. Aneliya Perkovska, who evidently collapsed from radiation sickness while carrying out her duties as Chairperson of the Prypyat Komsomol Committee,[44] stated that she was warned in hospital not to say that she was from Prypyat. It would be in her best interest to say that she was from Tahanrih or Kishinev. Perkovska ignored the advice because she felt that it was beneath her dignity to lie about her background. However, in the hospital during a meal period, two women from Tula and Kharkiv joined Perkovska at a table. Upon hearing that she was from Prypyat, "they immediately ran away." Instead, Perkovska was seated next to "comrades in misfortune," women from Chernihiv.[45]

Life on the temporary farms was harsh initially. Turning again to the Gorky collective farm in Makariv Raion, the conditions were said to be very difficult. Over a period of three hours on the first day after the arrival of the evacuees, 300 trucks delivered the livestock from the contaminated zone. Before dusk fell, the cows, calves, pigs and horses had to be unloaded, recounted, washed and driven to summer pens that had been constructed hastily for their arrival. Collective-farm chairman Mykola Tyapko stated that people were working for 12-14 hours, including women with children remaining at home. The process of counting heads of cattle naturally became more complicated in the twilight hours. The livestock themselves were in a state of shock, unused to the hours of jolting and the noise of engines. They "refused to come out of the trucks." The task therefore took much longer than anticipated as young women tried to drag 200-kilogram pigs from trucks until 4am.[46]

Work was found for the new settlers in Fasova and Lyudvynivka within three days, in contrast to the expected time lag of 1.5 months. The evacuees assisted in drying flax, weeding potatoes, beets, tomatoes, loading bales of hay onto tractors and arranging them under awnings. A small area of land was set aside for the people from Chernobyl in order that they "could actually see the fruits of their labor." Tyapko's task was to visit all the houses that had taken in settlers and to attend to their various needs. The requests of the settlers provide some idea of their immediate concerns. Thus one woman requested firewood. Another asked for a car so that she might visit her relatives in a neighboring raion. A third had left the documents for receiving her pension behind in the 30-kilometer zone.[47]

It was said that ignorance was a source of concern. Tyapko had to prepare himself for discussions on those themes that made people anxious. In the first few days he was asked most often about the radiation levels in the area, whether the No. 4 reactor unit was still emitting radioactive substances, what food was suitable for eating and which was unsafe. Then the evacuees began to ask the question that appears to have been uppermost in all their minds, namely: when will we be able to return to our native villages? According to another account, the question was an especially sensitive one, of moral and psychological concern. The reporter in question had visited people in both Chernobyl and the new homes for evacuees and had encountered two categories of people. First, there were those who had already returned to their homes. They considered themselves "the happiest people in the world." Second, there were the people for whom a return home had not been clarified. The "vast majority" in the second category, which must have been the largest of the two, considered themselves unhappy and they "complain bitterly about their fate."[48]

Even when the families had been provided with more permanent accommodation, the longing for the old homes apparently did not diminish. For example, 347 families from the Chernobyl zone villages Kopachi, Leliv, Chystohalivka and Bychky were accommodated in one of the first new villages to be built in Kiev Oblast, called Nedro, in August 1986. One family was described in some detail in a Kiev newspaper. The father was a locksmith and the mother a zootechnologist. They had a son and daughter aged 2 and 1 and lived in a three-room home. It was reported that the new homes were so pleasant that people from "outside" were trying to find jobs in Nedro in order that they might have such a home. A waiting list had been drawn up because all the job vacancies had been taken. Yet the featured family still missed their former home: "Such is human nature—the heart longs for native places."[49]

Because of the unhappiness about their situation, many people were permitted to return to their old homes, both officially and unofficially. Usually, the elderly were allowed to return home first. They went back to villages that officially were unfit for residence because of radiation fallout. In Lubyanka in Poliske Raion, for example, which in the spring of 1987 had not officially been repopulated, shiftworkers in the flood zone of the Illya River noted the lights in the houses of senior residents who had returned in the spring of 1987, and were now living happily in their old homes.[50]

Was it not dangerous for them to be there? The answer is yes, but that as the earlier example of the two elderly women demonstrates, it was virtually impossible for the government to stop them returning. Instead, the decision was made to permit them to return and to take dosimetric readings in the area. In addition, the return of the seniors had an important psychological effect on those who doubted that the area around the reactor was safe. K.T. Fursov, a Deputy Chairman of the Kiev Oblast Executive Committee, remarked that in his view conditions were safe, but the villages lacked various amenities and public services. Nonetheless "these people are coming back of their own volition. It's their home."[51]

For the majority of evacuees, the evacuation and subsequent period was a time of great hardship. Clearly the villagers had developed a great attachment for their native homes. It was undoubtedly easier for the predominantly young population of Prypyat to adapt to new conditions than it was for the local farming communities. As will be shown, however, the Prypyat residents had the prospect of being reaccommodated in the area. Those involved in work at the plant were engaged in shift work within the 30-kilometer zone. Some had never left the zone. For the farmers and their families, however, there were no clear prospects for returning home to many of the villages, especially if those villages, like Leliv and Kopachi mentioned above, were within 10-15 kilometers of the exploded reactor. For them, Chernobyl was a miserable and harrowing experience, a complete upheaval of their daily lives.

Heroes

The literature that has appeared on Chernobyl in the USSR is notable for the way it divides the actors in the scenes into heroes and villains. The accident was treated and perceived as a test of people's mettle. Some had passed the test, while others had failed abjectly. Because of the narrowness of this interpretation of individual actions, no clear picture emerges of the heroes. The villains, generally, committed certain actions that enabled them to be lumped into one category. For example, they may have run away after the accident, become intoxicated, stolen property and were thus condemned. But the heroes' personalities rarely emerge with normal human failings as they are elevated to almost mythical status.

In some cases, such as that of the Chernobyl firemen, the laudations appear justified. The firemen who fought the fire in the fourth reactor unit performed a heroic, selfless task. Yet the world's worst nuclear accident should perhaps not be seen as a real test of people's character. One can bear in mind, for example, that there were those whose first reaction was to flee from the scene to escape what at the time appeared to be certain death. Later many returned and worked within the 30-kilometer zone. The fact is not mentioned specifically in Soviet accounts, but can be deduced from piecing them together. There was the case of a doctor who volunteered to work in the 30-kilometer zone and then found himself somehow paralyzed and unable to perform his duties. Was he a hero or a villain? Had he failed a test, even though he had left his normal duties and traveled to the Chernobyl area? It seems that the gray zone is larger than that of the black and white, which raises the question why Chernobyl personalities have been divided up in this way.

The answer appears to lie in the myth of Chernobyl. The tag is somewhat simplistic in that Chernobyl was a serious battle against man-made elements, it was a destruction of the natural environment. But in the Soviet view, it was first and foremost a victory, a story with an ending, and an ending that was triumphant. Not all Soviet writers agree with this interpretation. Oles Honchar of the Ukrainian Writers Union, for example, maintained that the authorities had not learned lessons from Chernobyl and that an international conference was the only way to raise all the questions necessary about the accident and its consequences.[52] But the Soviet leadership and the Soviet Academy of Sciences have not varied in their interpretation of Chernobyl other than to play down its impact. And the official story requires its heroes and villains.

The role of the Chernobyl firemen has been recognized both inside and outside the Soviet Union. The first two firemen on the scene were Volodymyr Pravyk and Viktor Kibenok. In August 1986, a memorial was erected at the Uritsky Firefighting Institute in Cherkassy (Ukraine), where both Pravyk and Kibenok had studied. In the village of Boro-dyanka, the home of Pravyk's mother, located 80 kilometers south of the nuclear power plant, a street was named after Pravyk.[53] In the Fall of 1986, a ceremony was held at fire section No. 17 from which Pravyk and Kibenok's squad set out, and a speech was given by Volodymyr Skopenko, Extraordinary Ambassador and plenipotentiary of Ukraine to the United Nations. The section was presented with a marble plaque from the firefighters of the United States and Canada, which stated:

Firemen of Chernobyl! You stepped into the fire, saving your children and ours. We will never forget your courage.[54]

On November 5, 1986, Kibenok and Pravyk had the title of Hero of the Soviet Union conferred on them posthumously "for their for-titude, heroism and selfless actions displayed in dealing with the acci-dent at Chernobyl AES [atomic energy station]." Their relatives and other firemen assembled at the Presidium of the Supreme Soviet of the USSR in Moscow, and the awards were presented to their mothers, Irina Iosifovna Pravyk and Natalya Ivanovna Kibenok by President of the Supreme Soviet of the Ukrainian SSR, Valentina Shevchenko.[55] Another monument is being erected in honor of the firemen in the new city for Chernobyl plant operatives, Slavutych, in Chernihiv Oblast, replete with an "eternal flame." Near the monument there is to be a museum that will feature displays and photographs of the firemen and biographical accounts of their lives.[56]

The best known of the Chernobyl firemen was Major Leonid Telyat-nikov, the former station fire chief at the Chernobyl plant. Telyatnikov had lost all his hair after the accident and received a very high dose of radiation. Yet miraculously he survived, one of the few firefighters involved in the heart of the struggle against the fire who did not suc-cumb to burns or radiation sickness. From May until September he remained in hospital. Until July he was not even aware that his col-leagues had died during the struggle against the atom. Telyatnikov became a Hero of the Soviet Union shortly afterward and his picture frequently appeared in the Soviet press. In early March 1987, Telyat-nikov was invited to the United Kingdom by S.J. Duranty, the Presi-dent of the British Fire Services Association. Telyatnikov and A.K.

Mikeev, head of the Chief Administration of Fire Protection of the Ministry of Internal Affairs of the USSR, were received by British Prime Minister, Margaret Thatcher.[57]

Telyatnikov was awarded a gold medal for valor by the London *Daily Star*, and his photograph appeared on the front page of that newspaper, much to the delight of the Kiev newspaper, *Pravda Ukrainy*. A tribute to Telyatnikov and his colleagues was delivered by Ken Cameron, General Secretary of the British Firemen's Union, which (as translated from the Russian version that appeared in *Pravda Ukrainy*) went as follows:

> The heroic deed of Leonid and other Soviet firemen will always remain in the memory of mankind. As a man in this same profession, I can very well imagine what it cost—to work in such temperatures when you are literally swimming inside your protective uniform. Our work generally has risks, but what was achieved in the first hours after the tragedy in Chernobyl cannot be called anything but a heroic deed. And we, English firemen, are proud of what our Soviet comrades did.[58]

The tributes continued in 1987. A permanent exhibition to the Chernobyl firemen opened at the assembly hall of Schenectady, New York in April 1987, with pictures of the first victims of the accident: M.V. Vashchuk, V.I. Ihnatenko, Kibenok, Pravyk, N.I. Tytenok and V.I. Tyshchura. According to TASS, the Schenectady firemen had not swallowed "the bait of malicious anti-Soviet propaganda" being kindled in the United States about the Chernobyl accident.[59] In May, a museum called "In Memory of Courage and Glory" opened in Kiev which consists of three main rooms. The first room is dedicated to the memory of those "who died on duty," that is, the Chernobyl firemen; the second displays the equipment used by the firemen during the accident; and the third examines firefighting affairs today and how fire safety at the Chernobyl nuclear power plant has been improved since the disaster.[60] In East and West, there appears to be unanimity about the valor of the firemen and their role in preventing what could have been a much worse disaster.[61]

The firemen themselves, however, adhered to the official line that compared their actions during the fire in the reactor building to the courage of Soviet soldiers during the war against Nazi Germany, or else they would comment that Chernobyl provided a warning lesson about the dangers of nuclear war, following the example of Mikhail

Gorbachev during his television broadcast about Chernobyl on May 14, 1986. Asked by foreign journalists whether they were aware of the mortal danger to which they were exposing themselves in the early hours of April 26, 1986, V.F. Melnikov, the former Komsomol organizer in the fire section responded:

Of course we knew of the danger! Earlier we often pondered over the source from which the heroes of the Great Patriotic War drew their emotional strength, to cover with their bodies the gun-ports of hostile pillboxes, to go out into open terrain, not to give in under duress in the Gestapo torture chambers...and now the experience of Chernobyl shows that our generation too can, in the hour of need, act courageously.[62]

The anti-nuclear war message appeared again in yet another ceremony on behalf of Chernobyl firemen, held in Ceraso, Italy. It began in early July, when lesser known firemen were presented with the Golden Seagull prize, which is given to those who make major contributions to ecology by the Ceraso City Council, and is apparently judged by a private group of Italian scientists. Initial reports about the visit spoke of four firemen, but finally three were given permission to go to Italy to accept the award: Ivan Kotsyura, Vyacheslav Rubzov and Oleksandr Yefimenko.[63]

A month later what was described as a "massive meeting of Italian-Soviet friendship" was held in Ceraso. Evidently in July, the city authorities of Ceraso had sent a message of "peace and friendship" to General Secretary Gorbachev, and the August meeting was held in order to read out Gorbachev's reply. The local branch of the Italian Communist Party disseminated a circular which stated that:

The "Golden Seagull" Prize, founded by the Ceraso city council and given to the Soviet firemen, presents itself as our message of peace to the people of the USSR. The accident at the Chernobyl nuclear power plant forced everyone to think about the need to strengthen cooperation between countries in the sphere of scientific research.[64]

The heroism of the firemen, then, became politicized in that it was used to promote ties between Moscow and a branch of a European Communist Party. The irony is that it was the Prypyat branch of the Communist Party of Ukraine that had demoted Kovalevska for warning the party about the dangerous state of affairs at the Chernobyl plant. Again, the emphasis on heroes—and heroes they were—clouded past developments and solidified the myth of Chernobyl.

While the firemen were the most prominent in the heroes category, others also received accolades for somewhat less sensational but nevertheless hazardous work. By April 1987, over 700 people had received orders and medals of the USSR, and more than 600 had been presented with Certificates of Honor and Certificates of the Supreme Soviet of the Ukrainian SSR.[65] We will provide a brief survey of some of those who were featured in the Soviet press and in Soviet journals.

Two heroes were the brothers Senior Lieutenant Nikolai Uksukov and Lieutenant Sergeant Sergei Uksukov of the sections of chemical defense who were playing an important role in cleaning up after the accident. Both had completed the Tambov Supreme Military Command Red Banner School of Chemical Defense and had become commanders of subdivisions and specialists in the field. It was reported that they led by example, and took on the most dangerous work themselves before allowing their subordinates to follow them. Thus it was the officers who headed the patrols or led aerial reconnaissance. They themselves carried out the dosimetric checks. How had they learned such practices? From the example of Soviet soldiers at the front in the Great Patriotic War.

> It always happened that way; in the moment of danger—and here in Chernobyl this danger was fatal—Soviet warriors demonstrated the best of character traits: bravery, courage, selflessness, nobility, readiness, there was no reluctance about coming to help during the tragedy.[66]

There were also heroines of Chernobyl. They included two young women, Lyuba Vasilevskaya and Olha Snehir, who were operating a traveling library for those displaced by the accident. Vasilevskaya was a laureate of the Ukrainian and a graduate of the All-Union ballroom dancing competition, and had worked in the Prypyat House of Culture as an instructress in the art department, while Olha had been a choreographer in the same department. By October 1986, the families of both had been widely scattered, so Vasilevskaya had gathered some 9,000 books and moved to Bilyi Paroplav (White Steamship), the ship that was accommodating the building workers from the Chernobyl plant while a new village was constructed for them at Zelenyi Mys, on the Kiev Reservoir. Bilyi Paroplav is located on the periphery of the 30-kilometer zone. After four shifts at the steamship, the two women were to go to Sochi on the Black Sea for a vacation.[67]

Another woman ran a gas station at the entrance to the special zone. Buses came past daily taking people from Zelenyi Mys and Bilyi Paroplav to the Chernobyl nuclear power plant. Tatyana Izyumskaya, a young, blonde woman was known to the Chernobyl workers as "Queen of the gas pump." She was obliged to work very long hours. Also noted for their service to workers in the special zone, but in a very different respect, were Volodymyr Usenko and Valentyn Manayenko, the directors of a combined wind orchestra of 30 musicians. Their working day lasted from 6am to 10pm and sometimes even longer as they played 7-8 concerts daily to various audiences in the zone. They performed at the fourth reactor unit on the occasion of the sealing of the exploded reactor inside its sarcophagus.

In early January 1987, the Ukrainian youth newspaper, *Molod Ukrainy*, featured Aneliya Perkovska, a former schoolteacher who had studied mathematics at Kiev University before becoming the Secretary of Prypyat City Komsomol organization. The article described how she went to the local railroad station and then continues:

> Her roots are at another station...the name of which echoes in the hearts of millions of people with alarm and pain in all corners of the country, in the entire world. Chornobyl atomic energy station. Prypyat. Such is the station's address where you yourself, the Secretary of the Prypyat City Komsomol organization, Aneliya Perkovska, discovered a new formula—the formula of people's character. [She] did not spare either force or energy, or her soul or her heart when she came face-to-face with people's misfortune.[68]

Perkovska had reportedly seen the crimson glow in the sky during the night of April 25-26. Her colleagues were already at the scene. The on-duty dosimetrist, Mykola Horbachenko, for example, "forgetting about the terrible danger," had plunged into the flames to carry out victims, three times, five times, and ten times, when "all around him was collapsing" and the dosimeter had simply gone off the scale. Finally he was taken out by friends, staggering from the effects of a massive dose of radiation. The Komsomol member of the station bureau, electrician Serhii Kozachuk, also was said to have risked his life by remaining at his post in the switchboard apparatus, refusing requests that he leave and go into a safe zone.

Perkovska, as Secretary of the City Komsomol Committee, had the job of going around schools, kindergartens and children's playgrounds in the residential quarters where children were outdoors. The source does not make it clear when exactly she carried out this task. It seems more likely in view of other evidence that the children were warned of the dangers on April 27 rather than April 26. The implication here, however, given the chronological account, is that it took place on the morning after the disaster. Also, Perkovska went through thousands of Komsomol documents, especially accounting leaflets, in preparation for the evacuation. She worked all night on April 26-27 so that the entire collection of documents could be moved.

On the following morning, Perkovska organized volunteers to dig the sand that was eventually dropped on the reactor crater to help quash the graphite fire. She also kept a record of the amount of sand that was used. It was reported that her work continued for three days and nights without a pause. She was responsible for the evacuation of 5 sections of Prypyat, ensuring that pregnant women and children came first. Problems emerged because many of the evacuees became separated from their families. Perkovska had to help them move into their new localities and to help arrange employment for them. Perkovska alone, according to the account, remained in Prypyat after the evacuation, "like a captain on his ship." Eventually she was hospitalized. Subsequently, Perkovska worked in Zelenyi Mys and Bilyi Paroplav before taking a job in the Kiev Oblast Komsomol Committee as an instructor of young people in a special school. Even then, it was stated, she frequently returned to Chernobyl.

The story illustrates not only the bravery and fortitude of a young woman, but also the paucity of personnel dealing with the intricacies of the disaster. In the first days, the bulk of the work was clearly carried out by a few people. Why, for example, would a Komsomol Secretary carry out such a heavy workload rather than the Secretaries of the local city party organization? Why would Perkovska receive no relief after working for three days without sleep? Conditions were far from normal, but it seems plausible that many of those who should have been responsible for the work had left the area, among the tens of party members who remained unaccounted for months after the accident. This is not to belittle Perkovska's work, but rather to put it into perspective. Because she performed so many vital tasks, she stands out as one of the heroines of Chernobyl.

It should be added that many heroes of Chernobyl did not receive accolades, nor was their work ever publicized. Those involved in some of the worst work of the cleanup campaign received surprisingly little publicity. One reason may have been, as will be shown below (Chapter 5), that they were in the zone against their wishes, disgruntled and morose. Yet they outnumbered the official heroes who have been toasted by the Soviet authorities.

Villains

The evacuation of citizens from Prypyat and of residents of neighboring villages left a virtually deserted zone. In many cases, people did not have time to take their valuables with them to their new homes. Moreover, the unusual circumstances that arose after the accident encouraged a small wave of petty crime within the 30-kilometer zone. Periodically, reports of the misdeeds filtered into the official press in both Kiev and Moscow. Their importance should not be exaggerated, but on the other hand, it is undeniable that the events were a constant source of concern to the local police authorities.

On August 8, 1986, *Pravda* published an article on the "special attention zone" that was subsequently discussed at a meeting of the Collegium of the Ministry of Internal Affairs of the Ukrainian SSR, headed by Minister, V. Durdinets. Several leading officials in the ministry were fired or severely reprimanded for "inappropriate work for the protection of the personal property of citizens," including the police chief in the Chernobyl area, V. Skopych. The Collegium and the Kiev Oblast Executive Committee elaborated more rigorous control over law and order in the special zone. The entire perimeter of the zone was fortified. Militia workers were posted at each evacuated population point. Many of the homes in Prypyat and Chernobyl were linked up to a central protection board, while routine checks were carried out on agricultural structures.[69]

A number of people who had entered the special zone without authorization and had stolen property were convicted. At this time, the former residents were given the opportunity to visit their homes and apartments to check on valuables and examine security in the area. Yet the situation did not improve significantly, partly because some of the crimes seem to have been linked to evacuees who were allowed to return to the zone to check their property.

In mid-September 1986, more reports about theft and burglary in the 30-kilometer zone appeared in the Ukrainian press after complaints from evacuees about sluggish police work there. One evacuee had returned to the zone, ostensibly to check that his property was safe, and had been caught breaking the locks on neighbors' doors. Three maintenance workers at the Chernobyl nuclear power plant had also been discovered stealing shopping bags, a radio/cassette player, sacks of cigarets and candies from a Prypyat cafe.[70] Consequently, the Ukrainian Ministry of Internal Affairs began to step up security in the zone, using patrol cars around the clock and devices such as the gluing of strips of paper onto the doors of the deserted houses.

According to one of the police chiefs, there were more "misunderstandings" than thefts. For example, he stated, during the evacuation, a man took a valuable from his home and forgot to tell his wife. When she came back later to check their property, she was unable to find the valuable and called the militia. However, there was no misunderstanding about a car driver delivering supplies for the Raion Services Union who decided to bring alcohol into the zone—which is officially a "dry" area—and sell it for prices much higher than those set by the state. The man who interrogated the driver commented that vigilance was needed to ensure that people did not get rich off the misfortune of the "Chernobylites."[71]

In mid-October, a report declared that "lawbreakers" had turned up in the population points of the special zone. Militia had foiled an attempted theft in the village Novoshepelychi. A former resident of the village, who was referred to only by his surname Kukovets, had entered the village without permission as though concerned about his own property. Having examined his own property he entered that of his neighbor's and was caught by militia lieutenants A.I. Semenko and V.V. Poluyakh. Another case concerned Ostrovsky, a former cook in a Kiev restaurant who received products to organize hot food for people in the special zone, but sold the products for personal profit. Criminal proceedings were instituted against both the above.[72]

Why were the authorities so concerned about relative trifles? After all, Moscow itself is known for its "underground" black market trade of illicit goods. And were not the Chernobyl thefts minor affairs? In fact, it seems to have been the circumstances of the crime that infuriated the authorities. The petty criminals were taking advantage of a major human tragedy and deserved the most severe punishments. Writing in *Robitnycha hazeta*, V. Korolyuk commented that a few years earlier, he had regarded with indulgence and pity those people who risked their good name over small thefts. He had even written about them from a psychological perspective, trying not to traumatize the person. The word "thief" was replaced by the word "unfortunate." His views, however, had been changed by the outbreak of problems that had occurred in the special zone.[73]

There was the case of T. Kuzichkina who came to Borodyanka to work at a pioneer camp that was being used as a relaxation center for those recovering from the effects of the accident. While there, she had pilfered 10 tins of sprats, 6 jars of coffee and 20 tins of condensed milk (an important item). The total cost of the items amounted to 41 rubles 70 kopecks. A raion agricultural worker, N. Vlasenko, had tried to take out of a village general store a bottle of "Krasnaya dvozdika" shampoo and toilet soap worth 1.5 rubles. For the attempted theft he received a fine of 60 rubles. The same store was also the target of H. Dyabko, who tried to steal four cotton football shirts worth 19 rubles. His fine was 70 rubles.

Also in Borodyanka Raion, A. Ponomarenko, the Director of the brickworks of the "Novaya zhizn" collective farm had stolen 30 liters of benzine. A driver from Buchansk, V. Tarasov, who worked in Kladievo had tried to pilfer 20 square meters of lead from his factory. Finally, a worker from the Kiev Department of Mechanization, O. Panchik, had taken 40 kilograms of beet fodder from the "Zarya" collective farm in the village Yanova Buda, Borodyanka Raion. What incites these people?, asked Koroluyk angrily. In his view, the answer to such transgressions lay in proper educational work. The brickworks director Ponomarenko was a member of the Komsomol, and a "serious discussion" had taken place in the collective about his actions. But one should call things by their proper names, Korolyuk pointed out, "Thieves are thieves."

There were other misdemeanors beside theft. In late September 1986, certain ministries and departments were berated for failing to ensure the safety of highway traffic in the zone around the nuclear power plant. Following an article that appeared in a Ukrainian newspaper, over 60 drivers had been detained in the area of the reactor for drunk driving, despite the fact that no alcohol was permitted in the 30-kilometer zone. In October, 12 officials were admonished for sending out drivers who were not sober, including mechanics who worked for the Ministry of Power and Electrification of the USSR and for the Ministry of Motor Transport of the USSR. The Section Director of the association "Yuzhatomenergostroitrans" (Southern Energy Transport Construction), A.P. Kostyuchenko, was dismissed from his job on the watch shift for arriving at his work station in an innebriated state.[74]

The interviews held with local residents by Yurii Shcherbak reveal many cases of people who were reluctant to work in the special zone. It was reported, for example, that a veteran worker from the Kiev metro, who had been recruited to work on digging the tunnel underneath the fourth reactor, arrived in Chernobyl and began spreading rumors that frightened Kiev workers were refusing to come to Chernobyl. Even in Kiev, it transpired, the man was something of a hypochondriac, and in Chernobyl also he felt sick at the slightest thing. But all his diagnoses "turned out to be inventions." As a result, the metro worker was expelled from the party. Similarly, a young electrician who was also a Komsomol member refused to go to the special zone where he was needed to replace those who had already received their maximum "quota" of radiation. He could not be persuaded to go and was likewise removed from the party.[75]

Cowardice at Chernobyl was invariably treated with scorn and contempt. Moreover, those people who had initially fled the scene and then returned were also condemned outright even though one could argue that they had recognized their mistakes. These persons included many of the section leaders of the nuclear power plant's Building Department. The Director of the department, V.T. Kizima, a veteran, had remained at his post when his subordinates ran away. It was said that the return of these people was having an adverse effect upon agitation work in the special zone:

> They are individuals. But it would not be right not to speak of them. The point is that they, having waited it out around the corner, returned. Some became leaders again. And whereas earlier, during meetings, they turned away their eyes in embarrassment, now they speak out, recall the human factor, train people. And at best, the workers do not apprehend them. Therefore it is difficult to carry out agitation work in such conditions.[76]

Plainly, those workers who had remained on the scene resented the fact that their superiors who had not were allowed to return and take their old jobs.

The "villains" arose from the unusual situation of Chernobyl. In some cases, genuine criminals appear to have arrived in the area for easy pickings. But for the most part, the people concerned came from within the communities affected by the accident. The period of key concern seems to have been between August and October 1986, a time when the main danger of radioactive fallout had passed, and the thieves felt that they could take a risk of visiting the zone. Initially, the authorities were unprepared for an outbreak of petty crimes or for the bouts of drunkenness that ensued and which could perhaps be attributed to the pressures of working in a dangerous area. Chernobyl brought out all the sides of human nature, both good and bad. The authorities were concerned about the public's image of the disaster, however, and whereas they were prepared to highlight individual heroes and individual villains, all of whose names appeared in the press, many aspects of the darker side of the post-accident work were not brought to light. These occurred above all in the massive and controversial cleanup operation within the special zone.

5 The Special Zone

The 30-kilometer zone began as a somewhat arbitrary area that was considered sufficient for the purposes of evacuating the population. It was marked initially only by increased military patrols. By the Fall of 1986, however, its perimeters were more clearly marked. Special permission was required to enter and leave the zone. By the spring of 1987, the zone had become virtually a military camp. In turn, the special zone was divided into three subzones. The first was the region directly around the damaged reactor; the second was the immediate vicinity, in a radius of approximately 10 kilometers; and the third was the remainder of the zone. Conditions varied, but it seems that generally speaking, the further away from the reactor one worked, the less facilities and protective clothing there were for the workers.

Officially, conditions within the zone were quite strict. In order to enter, one had to pass through special sanitary posts. Upon leaving the zone, the workers were stopped at these same posts so that their clothing could be decontaminated. In theory, each worker in the zone was to be provided with an individual dosimeter, but in practice there were not enough dosimeters to go around. Before a work group went into the zone, dosimetrists preceded the workers, driving into the zone in "special" (perhaps hermetically sealed) cars. They would then define the safest route to the station and determine how long the work group was to stay in the area. Each worker could accumulate 25 roentgens of radiation after which the person was to be taken out of the zone and to jobs that excluded contact with additional doses of radiation.[1]

These levels of radiation for each individual worker depended upon the swift and regular replacement of the zone workers once the official maximum dosage had been attained. But in fact there were never enough voluntary workers available. Between May 1986 and March 1987, it was necessary to decontaminate 500 population centers, including 60,000 houses and other buildings.[2] It was a momentous task that required a rigid regimentation and at the same time the ignoring of some of the rules for work in the special zone. These contraventions will be dealt with below. Suffice it to say here that conditions in the zone were extremely severe.

In November 1986, control over most of the activities around the station was given to a newly established production association, called "Kombinat." The General Director of Kombinat, Yevhenii Ihnatenko, occupied the same building in the town of Chernobyl that had been used as the headquarters of the Soviet Government Commission immediately after the accident. Kombinat was responsible for the operation of the Chernobyl nuclear plant; for decontamination work in the 30-kilometer zone; for the supply of goods to Chernobyl and Zelenyi Mys; for the construction of Slavutych, the new city for plant operatives; for providing sanitary and hygienic safety for those working in the zone; and even for receiving foreign delegations to the station.[3] The main tasks in the zone by late 1986 were the completion of work on the sarcophagus; the decontamination of highly contaminated areas (the radioactive fallout was very uneven); the cleanup of the No. 3 reactor at the Chernobyl plant; and a gradual transfer of zone work from shift conditions to normal methods of work.

The Sarcophagus

The sarcophagus or concrete shell over the reactor began as a makeshift solution. It was essential to prevent the further escape of radioactive materials into the atmosphere. Thus the burning reactor had to be covered and sealed "for ever." Early reports suggested that the covering was perceived as something less permanent, however. Valerii Legasov hinted that the sarcophagus was perceived as the solution for "this generation," but that the next generation might find something better:

In principle it will last for hundreds of years, but our descendants may find ways and means of moving [the nuclear] waste elsewhere or of rendering it harmless.[4]

As work progressed, however, the Soviet authorities, although clearly frustrated by the pace of the work, grew more confident about the durability of the concrete shell. In the first weeks, work conditions were truly horrendous and the workers could only remain for seconds at a time (see below).

The timetable for the completion of the concrete shell over the damaged reactor was geared to the schedule for returning the first Chernobyl reactor into normal service. From the outset, October 1 was the deadline for both events. According to reports in the Soviet press from the summer of 1986, the reactor could not be restarted unless the sarcophagus had been completed. In practice, however, economic needs came uppermost. The first two reactors were started before the reactor shell had been properly covered. This was a startling example of Soviet priorities.

In mid-September 1986, G. Vedernikov, the Chairman of the Government Commission at that time, commented that work on the concrete casing being built around the No. 4 reactor should be completed by late September, and therefore the first two Chernobyl units would be back on-line by late October, or early November "at the very outside."[5] His remarks left no doubt that the two timetables were tightly linked. Thus he implied that if the covering had not been finished, then the plant could not be restarted. Earlier, *Pravda* had complained that a shortage of high-quality concrete was delaying work on the sarcophagus. Workers at local cement plants were hanging around with nothing to do.

When was the sarcophagus actually completed? The completion of the work was announced on several different occasions. There was some embarrassment about the fact that two Chernobyl reactors were re-attached to the national grid before work was finished on the shell. Consequently, it appears that premature announcements were made which gave every indication that work was ending on the sarcophagus. Thus on October 6, Radio Moscow announced that "The sarcophagus is basically completed," and praised the construction workers there for recognizing the importance of adhering to schedules. It had taken five months of work, declared the reporter on the spot.

In mid-November, about a week after Chernobyl-2 was returned to service on a test mode, the authorities announced that the No. 4 unit had "finally" been encased in a vault of reinforced concrete.[6] No mention was made about the erroneous earlier announcement. On November 27, 1986, the newspaper *Sotsialisticheskaya industriya* stated that the entombment had been "fully completed" a week earlier. The length of the process had been a result of the need to complete a second containment roof for safety purposes. Evidently a further week had transpired after the first announcement.

Yet the saga continued. On December 9, the Ukrainian newspaper, *Pravda Ukrainy*, stated that the final "capping" had been completed by December 7, and that Construction Department No. 605, which was responsible for building the sarcophagus from the outset, had corrected some "shortcomings" in the design. The collective, made up of 44 different nationality groups, had "discharged their duty with honor" said the newspaper. Finally, a week later, in an editorial, *Pravda* stated that "today" the complex of installations protecting the fourth reactor unit had been completed.[7] More than two months had elapsed since the first announcement about the work's completion. Either the authorities had underestimated the task and had had to revise some of their earlier work, or else they had misled the public about the progress of the work. In either event, they were anxious to assure the public and world opinion that the job was at an end.

Given the nature of the construction work, the delays were hardly surprising. The task was exceptionally complex. Engineering problems were solved "on the run." The data on which the structure was based were obtained from photographs, diagrams and video cameras because no one could get close enough to the reactor to take measurements or to get a proper view of how something looked. Specialists from the Construction Department No. 605 ran into frequent problems. The tomb was made up of huge blocks which were assembled some distance from the building of the fourth reactor. But the approach roads to the reactor were not designed to carry such loads and the drivers had to drive like stunt men to reach their destination. Having delivered the blocks, the next question was how to put them in place, since they each weighed several dozen tons. Eventually crane operators performed this particular function.[8]

Writing in *Izvestiya*, Andrei Pralnikov stated that there could be no ideal organization of work because the jobs were too complex. Sometimes when the builders, many of whom came from the Ministry of Installation and Special Construction Work of the USSR, had to choose from several alternative plans in a particular area, they chose one that was not the best and then had to make amendments. The periods of "idle time" that ensued from the shortage of concrete were especially problematic in view of the fact that the workers were accumulating roentgens while waiting for the concrete to arrive. Personnel were being "expended" because of the wait.[9]

Reportedly there was strict selection of candidates for the building of the sarcophagus. The volunteers had to have the highest of qualifications and excellent characters. For the most part, however, the builders, who came from various parts of the country, brought their own workers with them. The sarcophagus was divided into six "regions." The first region was led by A.N. Demidov, an engineer-builder from the Urals who brought along 225 men. The second, led by B. Drokin consisted entirely of workers from Krasnoyarsk. Eighteen temporary party groups were established in the area.[10] But where did the ultimate authority lie?

The Construction Department was answerable to the Government Commission until November 1986. The heads of the Commission rotated among Deputy Chairman of the USSR Council of Ministers. Initially, the Chairman was the Donetsk native, Borys Shcherbyna. He was followed in short order by I.S. Silayev, L.A. Voronin, Yu. D. Maslyukov, V.K. Gusev and G.G. Vedernikov. After the latter's period of office, Shcherbyna took over for a second time. An important role in work on the sarcophagus, but less clearly defined, was that of A.N. Usanov, Deputy Minister of Medium Machine Building of the USSR. The Ministry, which has a role in the production of plutonium and tritium for the nuclear weapons' program at Soviet nuclear plants, lost many of its leading personnel in the dismissals that followed the accident.[11] Usanov's apparent prominence at the No. 4 unit is a further indication that Chernobyl nuclear plant had a military function.

Demidov's group had to erect the so-called cascade, the protective wall of the encasement. The wall took one month to construct, and each of its blocks weighed "hundreds of tons." The wall was a means of protecting the men from high doses of radiation and also permitted them to work higher up the reactor. Automatic pumps were mounted on the walls that constantly emitted liquid concrete, which poured out of flexible hoses into the construction. Shcherbyna's assistant, Deputy Chairman of the Government Commission, Yurii K. Semenov, pointed out that the natural process of fuel decay was occurring.

To prevent the reactor from catching fire again by trapping too much heat, a ventilation system made up of heat-exchange devices was installed inside the sarcophagus. Gauges were put inside the structure to monitor the temperature of the nuclear fuel and a filtration system was inserted. The monitoring system was known as the "Shatro" complex and had been created in the Institute of Nuclear Research at the USSR Academy of Sciences.

By October, the sarcophagus had reached a height of 74 meters. Into its body had been poured 340,000 cubic meters of concrete, and about 3,000 tons of metal structures had reportedly been mounted. One aim was to build a dividing wall between reactor units 3 and 4 to isolate the undamaged section from radiation. The wall was built and the final stage was to pump in concrete. But the concrete poured out through cracks and there was heavy radiation leakage into the third reactor area. People were called in to patch up the leak from the other side of the wall, in the area of the damaged unit. The job required 150 people who each had to work for 5-minute periods in order that the leak was sealed as quickly as possible.[12] These types of difficulties were common and were largely responsible for the delay in completing the sarcophagus.

After the completion of the covering, the local population remained fearful of its safety. V.F. Shikalov of the Kurchatov Institute of Atomic Energy stated in mid-December 1986 that "there are many rumors that the reactor has begun to breathe," that "the wild beast is once again raging." These rumors continued in the spring, when some local residents believed that the reactor had begun to heat up again. As evidence for this was cited the fact that whereas the surrounding area and the roofs of houses were snow-covered, there was no snow on the sarcophagus.[13] Soviet scientists on the scene denied this story. According to the Soviet authorities, the temperature within the sarcophagus had been falling constantly, and no deviations in the natural physical processes had taken place since June 2. Thus the level of radioactivity within the sealed tomb in May 1987 was said to be only 40% of the level of June 1986.[14]

By the end of December, all Soviet sources concur that the work on the concrete shell had been completed. The No. 605 Construction Department reduced its personnel by 50% as workers were dispatched to carry out other duties concerned with the deactivation of equipment in the building of the No. 3 reactor unit, which was now considered the principal task. The closing stages of the construction work saw the building of a fence, almost 1 kilometer in length around the fourth unit, and the moving of uncontaminated soil to the immediate vicinity around the reactor.[15] Eight months after the disaster, a permanent monument had been created to the accident.

Valerii Legasov saw the sarcophagus as an important structure that would in the future be of inestimable value to scientists. A single diagnostic system, he noted, was supervising the temperature and radiation levels and ensuring that no radioactive particles passed through the filters.[16] A new reactor section was created early in 1987 to observe continuously the processes taking place inside the entombed reactor. Volodymyr Halkin, a former employee at the Kalinin nuclear power plant, who was appointed Deputy Chairman of the section, described his work to a *Pravda* correspondent in March 1987:

> ...the diagnostic complex is made up of 4 computers. Here, instruments are connected to numerous data-collectors inside the covering, which record the temperature, the neutron flow, the status of air, water, and gamma radiation....All this information is put into a general image. Dozens of points are monitored inside and outside the structure. There is also geodesic observation of the status of the covering; it weighs hundreds of thousands of tons. The observations are conducted by workers from the Novosibirsk Institute of Applied Geodesics.[17]

The account later described the system of air filtration. From within the sarcophagus, the air passed through 3 ventilation systems and then through a block of 20 covering filters. The latter had been designed to accommodate 200,000 cubic meters of air per hour, although as of March 1987, they were actually driving away about 80,000 cubic meters an hour.

In June 1987, when a group of foreign journalists visited the Chernobyl plant, several questions were raised about the sarcophagus and the safety of the exploded reactor. Mikhail Umanets, the Director of the Chernobyl plant, and his subordinates were confident that the danger had passed. They stated that the installed cooling system was no longer in use and that a natural cooling process was occurring. As the external structures were not experiencing any temperature increases as a result of the reactor, they were certain that the ferrous-concrete structure built around it "will last for hundreds of years." Scientists were monitoring all the physical reactions within the reactor remains, but thus far had not seen anything that gave cause for alarm. The current temperature of the reactor was said to be 95 degrees celsius, down from 140 degrees in November 1986.[18] The "shatro" had been built.

Yet the work had lasted two months longer than planned. It had been labor intensive, and important as the task was, it had delayed decontamination work in other areas. The sarcophagus had also a different dimension. The photographs that really brought home to world television viewers the enormity of Chernobyl were those taken from helicopters over the yawning orifice that had formerly been the reactor roof. Now this gap had been closed and sealed off. The residents who lived in the area, who, with few exceptions, were not permitted to get close to the No. 4 reactor, were less certain about the finality and definitiveness of the concrete shell. They were by now suspicious of many things, and a covering for the monster that lay within was just one of them. Surely, many felt, it could erupt again, unexpectedly and without warning.

Methods of Work in the Zone

The basic work carried out by the Kombinat production association consisted of three types of decontamination procedure. In the first place, the contaminated soil was simply removed. Over an area of more than 405 hectares, 30 centimeters of soil was taken from the surface and shoveled into steel containers, which were then reportedly driven to special radioactive waste dumps. The main dump for the waste was said to be near the No. 5 reactor unit, which had been close to completion at the time of the accident. The figure of 30 centimeters is by no means clear. Early reports about the cleanup process had stated that the workers' task was to remove 10 centimeters of topsoil. Yet the figure has in some later accounts risen as high as 50 centimeters.[19] Possibly the figure increased because the contamination of the topsoil turned out to be deeper than initially suspected.

A second task was to bathe the contaminated area with a special solution. In particular, buildings, roads, machinery and equipment were sprayed, and the contaminated water was, as far as possible, collected, purified and eventually buried. The effectiveness of this method evidently varied from area to area and depended upon the degree of pollution of a region. The third method, and the one which required the services of most of the cleanup workers, was to spray the ground surface with a solution that turned into a film which solidified and held all the radioactive particles, which were then picked up and taken away by trucks to be dumped at the waste site. If the soil was badly contaminated, then it had to be removed by bulldozers, poured into containers and "reliably isolated."[20]

The entire process was hindered by the churning up of dust by the trucks, but by the spring of 1987, a means of circumventing this problem had been found, which was to cover the sandy soil with a layer of latex solution and plant grass seeds in it. The layer of grass would then prevent wind erosion and the spreading of the dust. As contamination was also being dispersed by falling leaves, these also had to be gathered and taken to special locations.[21]

An example of the scale of the task was the work carried out in Khoiniki Raion, in the Gomel Oblast of Belorussia, in the northern part of the special zone. Here, the roads had been machine-washed by machines that had hermetically sealed cabins. It was necessary to replace the roofs of homes, barns, garages and verandas in a general area of almost 62,000 square meters. Workers constructed 45 kilometers of fences, built 32 kilometers of water pipes, and asphalted streets in an area of 66 kilometers. It was necessary to inspect and to close down many of the 5,159 wells in the raion.[22] On September 29, 1986, TASS reported that in Ukraine, about 7,000 square kilometers of land had been examined for radioactive contamination, and that a new stage of work was under way to discern the "dirty spots" outside the 30-kilometer zone.

Only three days later, TASS issued another report to say that decontamination work in the area of the nuclear plant had been basically completed, which could only have been a source of grim, ironic amusement to those in the special zone. In fact, as Vice-President of the Soviet Academy of Sciences Evgenii Velikhov indicated in his meeting with the U.S. Senate Labor and Human Resources Committee in Washington in January 1987, the completion of decontamination could take up to 30 years. He also stated that 500 villages and 60,000 buildings had been decontaminated.[23] The number of villages stated was much higher than encompassed by those within the 30-kilometer zone, which was around 100 (including 69 on the Ukrainian side). Evidently it included adjoining areas in Zhytomyr Oblast, Chernihiv Oblast, the Bryansk Oblast of the RSFSR and others. It should be emphasized that the majority of villages cleansed were outside the official zone.

Transport into and out of the zone came under the direction of the State Auto Inspectorate, known from its Ukrainian initials as DAI (Russian: GAI). The DAI was not unique to the zone, but in fact operates throughout the republic and the country. DAI "boxes" are a familiar sight in every Soviet city. In the special zone, however, the work of the DAI took on a new and vital importance. The work of the DAI was coordinated by the operative staff of the Ministry of Internal Affairs of the Ukrainian SSR, led, during the post-accident months, by V.I. Sherstyuk.[24]

The DAI posts within the zone were located in the most sensitive areas. There was, for example, a post close to Zelenyi Mys, the shift settlement for workers at the Chernobyl plant. This was known locally as a "customs" post because one of its chief functions was to ensure that the truck drivers' cargo did not include alcohol, and checking of the trucks' cargo was said to be "extremely necessary." Another DAI post stood at the entrance to the zone, and motorists of all descriptions had to pass through a barrier before they could enter. The work of the DAI in the period of the major post-accident cleanup was so varied that several examples need to be provided to illustrate its nature.

In a report of October 1986, a writer from a Soviet journal visited the zone and was escorted in the patrol car of the DAI dosimetric control section. He noted that the DAI workers in the zone were strictly volunteers who had been sent there through the personnel department of the DAI in Kiev. The two who escorted him were both residents of Ukraine: Sergei Trebukha from Yalta and Volodymyr Lennikov from Lutsk. The writer's first example of DAI work was a report on the car's radio that a dump truck had exceeded the speed limit in the zone. It turned out that the driver, new in the zone, had violated the speed limit in order to complete his assignment more quickly. The driver was given a warning. It was explained that each driver is permitted three warnings before being prohibited from driving in the zone.[25]

This sort of infringement was apparently very common. The Commandant of the DAI, militia Colonel O.P. Shepotko, who had been sent into the zone from Sumy Oblast, commented that because both the drivers and car inspectors worked directly in the area affected by high levels of radiation, they often tried to carry out their tasks as quickly as possible, and thus violations of road safety, especially speeding, were not uncommon. The drivers, he stated, were under psychological pressure. At the same time, they were not permitted to linger in the zone either. Road conditions were difficult. The roads had to be rinsed constantly and skidding occurred when vehicles braked suddenly. Because of the radioactive fallout at the sides of the road, it was not permitted to drive on the hard shoulder.[26] Small wonder then that, as noted earlier, drivers sometimes had to drive like stuntmen.

When road violations occurred, discipline was reported to be strict. Meetings were held, which all the drivers attended, and "Windows of the DAI" were set up, which publicized road transport incidents and, according to Soviet practice of exposing the guilty to the public, photographs of the transgressors were displayed. To the Western mind, such methods seem harsh, especially given the fact that the DAI workers were reportedly volunteers and yet were obliged to work long hours in extreme conditions.

The average DAI employee worked a 12-hour shift, wearing a protective mask, checking the unending lines of concrete trucks, cars and bus loads of workers being taken inside and outside the zone, both for cleanup work and work at the Chernobyl nuclear power plant. Traffic jams occurred regularly at peak periods. Immediately upon entering the zone, the DAI worker received a brief medical check, and every three days he was obliged to take a blood test. After the 12-hour day, he was given another checkup and then taken to the rest center in the village of Ivankiv, just south of Chernobyl Raion. How long the average DAI inspector endured such a routine is not made clear from Soviet sources. What is clear, however, is that many spent several months working in these conditions, separated from their families. One inspector, a militia Captain, H. Tymokhin, for example, had been working in an area directly adjoining the nuclear power plant building for three months by October 1986. Asked how he was feeling, Tymokhin maintained that "it is a sin to complain about one's health."[27]

In winter, conditions became even more severe. One fleet of buses was maintained at Leliv, close to the city of Chernobyl, well within the special zone. From here, it was only a short drive to the nuclear power plant. Here, under "field conditions," worked 37 people who had been sent from Kiev, including 3 fitters, an electric driver and bus drivers. During the winter period, it was considered risky to allow the engines of the buses to freeze over and hence every two hours they were warmed up for 10-minute periods. All night, two drivers patrolled the 32 buses constantly. In the morning, 27 drivers began taking plant operatives to and from the station, working in a single "chain," that is, the buses set off at intervals one after the other.[28]

The peak period lasted until 9am. Afterward, the drivers carried out the so-called "inter-peak" work, fulfilling diverse services. A period was set aside for repair work and then immediately after lunch, preparations were made for the second peak period. At 6pm, the buses were again lined up in the chain and made their way back to the nuclear power plant. Between 6 and 9pm there were regular services from the station. Severe punishments were imposed for falling behind timetables. Drivers on the routes also had to take along cables in case the buses broke down or they had an accident. The days were not without problems. "Stupid bungling" and "elementary mismanagement" had raised their heads, mainly because of overenthusiastic workers who did not follow regulations. Some workers tried to be "heroes" (in what way is not made apparent) causing "moral and physical losses" (accidents?) thereby provoking the irritation of their fellow workers.

A report in a Kiev newspaper about the DAI's work suggested that the staff there was almost overlooked by observers because the work they performed was relatively undramatic and routine. Justifiably, the newspaper set out to correct this image. There is little doubt that the DAI workers risked their lives during their time in the zone. After all, as noted, some were working outdoors, relatively close to the reactor, in the middle of winter and for long and arduous hours. Moreover, whereas the sarcophagus had been built by December 1986, and part of the area immediately around the reactor had been decontaminated, the work of the DAI was unceasing and indeed continues today.

Reflections on the Zone

Before examining the life of cleanup workers in the zone in detail, it is of interest to look at life there through the eyes not of experienced military commanders or civil defense workers, or even of the leaders of the Kombinat association, but rather from some of the young students who arrived from the Obninsk Institute of Atomic Energy, from Sverdlovsk and other areas. They carried out various tasks in the zone, most of which were outside the first zone around the reactor itself. Shortly after they arrived, their commander, Igor Anshigin, advised them to keep diaries about their activities in the zone. In late October 1986, the newspaper *Komsomolskaya pravda* published excerpts from some of these diaries, which were revealing in their candor and were apparently unedited. What follows are translations of some of the entries.[29]

ENTRY 1
I appeared in the detachment by chance. As they say, a man from the country, a stranger. The point is that they took only sixth-formers.[30] But I asked the commander anyway. I turned out to be the youngest. There are beautiful places here: a forest, river, field. The empty villages are striking. I will never forget my first trip to Pripyat. The open gate squeaks. Wind makes the dried-out linens flap during the day. There are scores of lone cats. They sit on the sidewalk and watch the cars go by. They literally understand what happened here....

ENTRY 2

This is the first time I've opened our notebook. I was in Pripyat together with the operators. They were getting their belongings. We finished and waited for the car. One of the fellows took an apple from his pocket, awkwardly tossed it in the air, but didn't catch it. He brushed off the dirt with the flap of his jacket, as if nothing were wrong, and into his mouth it went. "What are you doing?" I was astounded. "What's wrong?" he asked. I am staggered by the absence of elementary knowledge about radioactive contamination. And among whom? Among those who in their line of work should know this by heart. I think that carelessness and the lack of elementary knowledge is the basic reason that brought about the Chernobyl accident.

ENTRY 3

Today we went to take measurements. If you wake up in the early morning in any other city you will see: there is a light on and someone is stirring in the kitchen. Or you will hear how the first tram rings from the depot. Here in Pripyat, it's different. Emptiness. There are no people. Pripyat provides a sharp contrast with Chernobyl. There, in the city, there are many people around during the daytime. They live. Work.... Everyone knows Chernobyl. Only here did I truly find out about Pripyat.

ENTRY 4

The day began as usual. I received an assignment to go to Pripyat. They define what can be taken out for residents. The usual apartment. The occupant—a 20-year-old woman, who looks like a playful child.[31] Laughing, joking. This was before we had gone upstairs. But when we entered her apartment, her appearance changed immediately. The smile vanished. Her face became pinched. She aged ten years. Then tears gushed. She was not shy about crying in front of me. By accident, I happened to see a note: "Volodyushka! We were evacuated. I will try to get to Horodyshche by whatever route. Try to get in touch or I will go crazy. The milk burned. Get well. Kisses. Nadya and daughter." That evening I learned that this was the wife of a fireman who had perished. And when Nadya wrote the note, her husband was already dead....

ENTRY 5

The flow of people was running dry. I was tired. I could not see the divisions on the dosimeter anymore. It irritated me that there were so many cars, so many belongings, so many people. And the fact that it was night and we had to work. I went inside our little house. I sat on the bench. Leaned against the wall. Nodded off but didn't really sleep, yet half an hour later I came out absolutely refreshed. A bus rolled up. An old woman came out. She looked at me and started to poke something at me, clutched in her fist. I grasped her hand: on her palm was a crumpled ten with a frayed edge. I felt my face turning red. I began to refuse, but she said: "This is for you, for your work." I tried to convince her that I get money anyway, and for her, an old person, it's shameful to demean herself so, and that it's normal to treat old people with respect—but I did not convince her. She did not comprehend my words. I almost started to cry. But, you see, she certainly had lived through war, devastation, famine. Then, relations between people were more brotherly, mutual aid and support were more sharply developed. And then there was a mother of five children. I immediately saw that the children's clothing was "dirty." [i.e., contaminated—D.R.M.] She did not want to hand it over. I swore at her, demanded, but there was no way. In her eyes was confusion, hurt. The clothing itself she threw out, but the expression in her eyes completely unsettled me.

ENTRY 6

August 19. The weather is appalling. Here and there rain pours, a downpour broke out as the main flow of cars went through. Like someone had literally overturned a barrel of water. Just in time I climbed into the cabin of the vehicle. A deafening crack rang out, lightning flashed. The searchlights went out and I found myself in pitch darkness. I was instantly soaked. My clothing stuck to my body like plaster. I shivered. I ran into the duty room. I barely had warmed up. But cars were waiting....I put on protective clothing, hung a dosimeter around my neck and went out into the street. I felt like a half-stranger, a sort-of knight from the Middle Ages. I had concealed myself from the rain, but something else was lying in wait for me. A gray, unremarkable man, glancing around furtively, slipped out from around the corner. He grabbed my arm and said: "I will give you a case of vodka!"

"For what?" I did not understand at first. He pointed to the container. He wanted to rummage through the "dirty" articles and take things.

"What for?"

"Well, to sell them in the second-hand market, or at a commission store."

I was furious. I yelled all sorts of things. During that night, several more such types appproached me. Then I understood: we were standing in no man's land. On one side—the zone. On the other—you can walk or ride right up. I barely lasted until the end of the shift.

ENTRY 7

How many people walked by in front of me. Young, elderly. Men and women. Their misfortune is the same. But they are all different. The men smoke nervously. The women cry. Today I attended to a young fellow. One of his boxes was "dirtyish" on top. "Let's look inside," I ordered. He started to bring it closer and dropped it. It rang out like an elephant had entered a china shop [sic!]. It turned out there was a set of chinaware inside. The fellow spread his hands helplessly and said: "For good luck." And then he told me this story. "A man helped to build an atomic energy station abroad.[32] His wife stayed at home. She was just about to give birth. We had put a gift in storage long before. And suddenly he received a telegram: he has a son! He walked joyfully around the city,[33] and imagined: he will have two more—no, three sons, they will grow up, they will have their own children, and someday they will all gather together in his home. Then his glance fell on a shop window: there sat a luxurious dinner service set. Just right for that future reunion, he decided. So he went and bought it. It's no big deal," he waved his hand. Then he said: "...I won't tell my wife it was contaminated. I will say that I didn't break it by accident. I did it myself. Especially. For good luck."

ENTRY 8

These days, they say and write quite a lot that young specialists are not trained to make decisions, to take on responsibility. In my opinion, this is true. At least, before this month, before this job, I didn't feel the need for such actions. Here, for the first time, I sensed what responsibility is. You see, the dosimeter's readings are not always unanimous and clear, and people plead, cry, threaten. One must decide simply and alone. My relatives did not let me go into the detachment. Now I cannot imagine that I could not have ended up here. We have grown up.

ENTRY 9

I am Anatolii Samonov. I didn't bring my license with me for nothing. I drive a car here. I drove around with Kostya Bespalov, Sergei Yasinsky, and Misha Chepushtanov. We took specimens and analyzed changes in the radiation background. In the entire 30-kilometer zone. One cannot make one iota of error here.

ENTRY 10

We found out that in Zelenyi Mys, at the shift settlement, Leontiev was to perform. This was no good to me because I worked the night shift. The concert was set for 10.30pm. The boys were even somewhat offended. But what can you do, to be fair, a schedule is a schedule. And since the morning, a hurricane has started. Like leaves in the Fall, the branches in the trees were falling to the ground. It started to rain incessantly after supper. But they did not cancel the concert. I see: they had to set up the equipment, but the square was open....Leontiev's electrician was running around cursing. If he shuts down, all the equipment will burn out....The plant operatives returned to the camp and had supper, some of them waited. By 10, the square was packed: they sat under umbrellas, oilskins and simply raincoats. And then the lights went out. For an hour and a half they tried to find out what had gone wrong. It turned out that the gale had torn the wire, ripped down the support. They phoned "Kievenergo." There they replied "we can't fix it before morning." What was to be done? The Secretary of the nuclear plant's Komsomol Committee and I ran—the people sent us—to Valerii at the steamer,[34] to ask: would he sing? "I will," he said, "Give me some light." They started to pull a cable from the steamer. It didn't reach. By about 1am, it became clear that nothing would work. The next day he was to perform a concert in Chernobyl, and then he was leaving. We went one more time to Leontiev. And he was upset, he had tears in his eyes, honestly. "I will tell you what," he said, "Tomorrow, late at night, I will come here after the Chernobyl concert."

ENTRY 11

It's hot. It turns out that we don't always have torrential downpours or hurricanes blowing here. The severity of the first impression has already passed. We have already become used to the familiar walkway. The empty city and the overgrown streets don't surprise us. Nor the fact that no one picks the apples, apricots or plums. And we're used to roses, there are thousands of bushes here. And to people's faces under respirators. And even to the strip of burned out forest next to the station and to the "dirty" sections. Our days are similar from one to the next and very dull. But the work here cannot be a dull affair. The zone cannot become habitual.

ENTRY 12

My entry is certainly the last one in this notebook. Because our shift is over. Tomorrow we will fly home, to Sverdlovsk. The tickets are already in our hands. The baggage is assembled, But I suddenly, agonizingly, wanted to stay here. Is everything really finished? The trips into the 30-kilometer zone? The days when the analyzer overheated from uninterrupted use? Will we really not stand anymore at the dosimetric control? Our work is very necessary—this is what I came to understand here. For those who are liquidating the consequences of the accident. And for those who will work here afterward. I would like to remain here. But I must finish my education, earn my diploma and defend it. Maybe on assignment I will come here to Chernobyl nuclear power plant. There is work to be found here....

There is a naivety about some of the entries. These were essentially youngsters recording events that were unique to them. Despite the sentiments expressed in the final entry, the students were fortunate in that their stay in the special zone, like those of most of the civilian workers, was limited to 30-day periods or less if one worked in areas of especially high radiation.

The Cleanup Crews

Less than two weeks after the Chernobyl accident, the authorities realized that if the area was to be decontaminated quickly, then much greater reserves of cleanup labor were required. The civilian volunteers were not particularly well disciplined, they were costly in terms of wages, and even more important, they could in theory leave their work at any time. The coal miners, metro workers and others who attempted to pacify the smoldering reactor were not volunteers. They were ordered to the scene. Those who refused to go, as is made clear by the interviews of Shcherbak, suffered rigorous punishments. Party members, for example, were expelled from the party. Others were, at the least, exposed and named publicly in the press. Yet subsequent work crews appear to have been voluntary, as workers came from as far east as Sakhalin Island to participate for 30 days in the cleanup operation.[35]

Some of the worst work inside the zone was performed by student "volunteers." Many were training to be firemen or for work in the nuclear power industry. Cleaning up the roof of the reactor, for example, which was one of the single most hazardous jobs in the entire operation, was undertaken partly by students from firefighting schools, because of their experience at working at such heights.[36]

On May 6, military reservists arrived at Chernobyl from various parts of the Soviet Union. Their exact number has never been specified in Soviet sources. However, an admittedly crude estimate can be deduced by subtracting the figures provided on the number of workers in the Kombinat association—7,000—from the number of cleanup workers provided by Velikhov in Washington in January 1987—40-50,000—to arrive at a figure of between 33,000 and 43,000.[37] The figure is staggering in that it would be about one-third of the number of Soviet troops inside Afghanistan during the same period. Uncorroborated reports that have emanated from Estonians in Sweden have suggested that approximately 4,000 Estonian troops were called up (the Estonian case is dealt with separately below), and given the size of the Estonian SSR, it seems plausible that if this were the case, other republics—Ukraine and Belorussia in particular—could be expected to send considerably higher numbers than Estonia. While the figures are uncertain, therefore, they do not seem improbable.

In April 1987, the cleanup work was commended by Hans Blix, Director-General of the International Atomic Energy Agency (IAEA), who commented that although he had not always considered himself an advocate of centralized planning, in this particular instance, it had proved its advantages.[38] In short, he believed that military-type discipline was essential to work in the zone. Captain Nikolai Makov, who led dosimetrists into various areas ahead of the workers, had served in Afghanistan. It was said that just as he would be the first to enter the battle against rebel gunmen there, so also at Chernobyl he led the troops in the fight against radiation contamination.[39] The description is dramatic, but it indicates the nature of the task. Further, as Soviet sources have often pointed out, at least in Afghanistan, the *mujahideen* sometimes missed their target, whereas radiation always found its mark.

The basic problems for all zone workers were threefold. First, they lacked knowledge about the actual situation within the zone. Second, even when they had been made aware of the area they were entering, there were hotspots of radiation in diverse areas that had not been identified. Third, and, it later transpired, the most significant difficulty, was their lack of experience at working under such conditions.

The crews had diverse tasks. Whereas some worked directly on the damaged reactor, Zone 1, others were in relatively remote areas on the periphery of the zone. Accounts that appeared in late June and July 1986 in the Latvian newspaper, *Cina*, were described by a Western authority on that republic. Evidently the Latvians had a campsite about 20 kilometers from the entrance of the zone. The men were said to vary in age from their twenties to their forties, although a year later, Oleksandr Kovalenko of the Kombinat association maintained that all the soldiers were over 30 years of age.[40] The Latvians were given responsibility for checking the radiation levels of drivers and vehicles leaving the zone, and for hosing down the vehicles and carrying out various measures for decontamination if the drivers and vehicles were over the maximum limits. After four such attempts to bring down levels had failed, then the vehicles remained in the zone. Camp life for the Latvians was said to be monotonous and there were some "disciplinary problems."[41]

The situation was evidently complicated by a fire that broke out several floors above the No. 4 reactor on May 23, 1986 (it was revealed by the Soviets only on August 20, 1986). Four fire engines fought the blaze during the night after it had been discovered by firemen who were pumping water out of the reactor building. The radiation detectors reportedly went off the scale as the firemen tried to get close to the source of the fire, and thus the firemen could only work for a few seconds at a time, pouring water on the blaze and then sheltering behind a concrete slab. The fire continued even into the following shift and for most of the next day.[42] There were no explanations as to why it had occurred or whether any casualties had resulted.

On the other hand, it has been stated elsewhere that the cleanup crews on the damaged reactor area were working not only in conditions of very high radiation background, but also in intense heat, with the temperature reaching 60 degrees celsius.[43] In these conditions, some of the workers panicked. Others, however, showed excessive bravado. They would "jump into the area" and by remaining much longer than the authorized time, "got injured by acute percentages" of radiation.[44] On various occasions, the workers were limited to only seconds or minutes on particular tasks. But over the summer of 1986, conditions in the zone became worse rather than better, partly because of the dearth of experienced personnel there. On September 2, *Pravda* commented that the work had entered a sensitive stage and that consequently it would be better if the personnel in the zone remained unchanged.

Various cases were cited where individual workers remained for longer than the alloted timespan in an area. For example, one newspaper noted that Viktor Zavedy, the team leader of concrete pump operators at Ignalina nuclear power plant in Lithuania, had come to Chernobyl, and that he "could work for 2 hours where presence is limited to no more than 30 minutes."[45] One cleanup worker, who was decontaminating an area immediately adjacent to the fourth reactor, had been there for almost six months. The story of his activities was reported in a manner that suggests that such lengthy periods in the contaminated area were not unusual.[46] A report on decontamination work being carried out by soldiers in October 1986, revealed that the men would often work a little extra time on decontamination, even though they were not paid for the work. Work breaks were reduced. Although in theory the men were permitted two days of rest per month from their 12-14 hour working days, even these periods were often "shortened" because the work was not being carried out quickly enough.[47]

In addition, the cleanup crews were often without protective clothing and shower facilities. A trade-union conference that was held in Chernobyl in October 1986 noted that insufficient funds had been provided to obtain protective garments for the cleanup crews. Speakers at the meeting put the blame on the trade-union, which was said to have the funds, but was inept in organizing their distribution. The point remains, however, that at least a portion of those involved in cleanup work did not have protective clothing (the lack of protective clothing was also one of the points raised by Gubarev in his play Sarcophagus). The same meeting also revealed that some of the men working near Leliv, just a few kilometers from the No. 4 reactor, had nowhere to take a shower after the day's work ended. There were shower baths at the relaxation center, but in order to get there, the men had to walk for two hours, in clothing caked with mud, according to one of the workers, Sergei Khoroshilov.[48]

There were other complaints. Workers who had been in the zone for more than five months still had nowhere to live. They remained in summer accommodation (often in tents) in the late-Fall period. They had been promised and even assigned housing, but this housing had failed to materialize. Although medical checkups were an important facet of the work, medical facilities had been neglected. Immediately after the accident, it was reported, the medical-sanitation point No. 126, which looked after the needs of nuclear power plant workers and residents of Prypyat, was almost 1,000 strong. Subsequently, however, it had been partially disbanded. Most of the specialists had left the area. Those that remained were moved to the village of Teteriv, where their "opportunities [for carrying out their practice] are very restricted." The implication is that medical care for those in the zone was being reduced.

While the facilities for and attention to the cleanup crews were being neglected, the workdays were severe tests for those involved. For example, the covering of the machine room of the No. 4 reactor unit was said to be the most dangerous place in the nuclear plant zone. High levels of radiation had been created by scattered tidbits of radioactive graphite, ferrous concrete and radioactive dust. The active operations area was 1,500 square meters in size. The fire service was called in by the Government Commission, and it relied on the work of students from the Kharkiv and Lviv fire-technical schools: Klymchyk, Horodenko, Kushkov, Hadzhyev, Dzyuba and Polochuk. At one stage in the work, a powerful source of radiation was discovered at the "third site," from underneath a layer of graphite.[49]

Similar stories were reported in the Latvian press, following the arrival of Latvian reporters at Chernobyl in late October 1986. Some of the Latvians were involved in work on the 20-story-high roof. Clothed in lead-lined protective suits, they ran up the flights of stairs, dug up a shovelful of radioactive debris, threw it through a gap back into the damaged reactor, and then ran back down the stairs. The intense radiation levels made it impossible to do more than this basic task. The Latvian Komsomol newspaper stated that everyone who was working on the cleanup operation was risking his health—evidently a dig at those spokespersons who had claimed that there were no dangers to the health of the cleanup crews—but that the workers were aware of the risks they were taking.[50]

One of the Latvian reporters, Adris Sprogis of the Latvian Writers' Association, commented that:

> It cannot be denied that for some the daily pressures at Chernobyl proved to be too heavy. For some, it was too much physically. For others, their nerves gave out. The sick were transported to a hospital in Kiev or sent home.[51]

The same report revealed that some of the workers tried to devise ways of leaving the zone before their officially allotted time was up. Sprogis maintained that some would volunteer for dangerous tasks in order to shorten their stay, while others would not show up for the job (one was said to be asleep in the bus that had brought the workers to the worksite). An account in a Ukrainian source also mentioned that there were cases of workers "cheating" in order to be relieved for a couple of days at the health center (while remaining on full pay). The most common method was said to be leaving one's personal dosimeter in an area of high radiation, then collecting the apparatus again and visiting the radiation monitoring service. The ruse was said to be successful until at length the service workers realized that in the areas where the men were working, radiation levels were lower than were being registered on the dosimeters.[52]

The reluctance of some of the cleanup workers to carry out their jobs, under the circumstances of dangerous and often deplorable work conditions is understandable. But it only put more pressure on those who remained behind. Even shift supervisors at the Chernobyl plant were dragged into the cleanup work, such as Valerii Zakharov, the supervisor at the Chernobyl-1 unit, who had been scheduled to head the shift on April 27, 1986, but instead had to work for a time on the cleanup (by December 1986, he had returned to his former job).[53] On numerous occasions, workers went over the official limits for radiation accumulation.[54]

What were these limits? On May 14, 1987, the Ukrainian newspaper, *Radyanska Ukraina*, acknowledged that the acceptable radiation dose for workers in the nuclear power industry was 5 ber (body-equivalent roentgen, or rems) annually. Technicians who came to Chernobyl from other nuclear power plants or who worked normally at the Chernobyl-1 and Chernobyl-2 reactors, were permitted to work only for a month, or until they received 2 rems. The lower dose was instituted because these people worked for a living in areas that had the possibility of increased levels of radiation background. From the summer of 1986 to December 31, 1986, the official maximum level for the military reservists was 25 rems,[55] but in practice, there is no doubt that this level was frequently, if not regularly, exceeded.

This becomes evident from various reports. We have seen above how some workers stayed in "hot" areas from bravado. Others remained because their replacements had not arrived in time,[56] or because the authorities did not want to replace those workers who had now learned the intricacies of decontamination work. The radiation levels around the reactor were up to 1 rem per hour in May 1986. Thus, in theory, many of the workers, especially those close to the exploded reactor, could have accumulated the maximum levels of radiation within two days. Yet it is known that a considerable number of workers stayed in the zone for six months at a time, and that even their two days per month break was often curtailed in the interests of speeding up the process. Even Anatolii Romanenko, the Minister of Health Protection of the Ukrainian SSR, who has hardly ever acknowledged that any Chernobyl workers or evacuees have been in any danger as a result of the disaster, admitted that:

> Of course, those who selflessly worked in the zone of the fourth block during the first few hours and days after the accident, did, unfortunately, receive a higher dose [of radiation].[57]

Inexperience, ignorance of conditions and other factors clearly played a role in the grim scenario being enacted. Yet the authorities also appeared willing to take certain risks for the achievement of longterm goals. Thus if a reservist worked with military discipline and took orders from above, he would be more likely to complete the task quickly and efficiently than a new person brought in from the outside, perhaps a civilian. Why not, then, keep the soldier on the same job, thereby ensuring the more rapid decontamination of the territory and the ultimate security of thousands of people? As Novosti stated:

> Nuclear plant accidents always cause extreme situations. Chernobyl was no exception. Chernobyl forced people curbing it to work in increased radiation. There were, alas, limitations to the use of robots and radio-controlled devices.[58]

In other words, the cleanup workers were performing tasks that preferably would have been undertaken by machines because they involved work in high radiation conditions. But were there human rights' violations during the cleanup work process? The verdict remains undecided, largely as a result of the unanswered questions in the Estonian reservists affair.

The Estonian Affair

In August 1986, a series of articles appeared in the Estonian Komsomol newspaper, *Noorte haal*, which led to some serious questions being raised in the West about the nature of the entire cleanup operation. The articles, which were uncovered and translated by an Estonian-speaking American scholar, Toomas Ilves, were written by Tonis

184 The Social Impact of the Chernobyl Disaster

Avikson, a foreign affairs correspondent for the Tartu newspaper, *Edasi*, who arrived in the Chernobyl zone at the start of the 1986 summer, and talked to Estonian workers at the worksite. Despite the frankness of the article—it went beyond anything that had appeared hitherto in the Ukrainian and Belorussian press—Avikson's own views were not always in sympathy with those of his interviewees. He revealed the nature of the discontent without necessarily agreeing with the complaints. Yet this added to the articles' authenticity rather than reduced it. On the other hand, the articles leave no doubt that Avikson was taken aback, if not shocked, by what he discovered in the zone from his conversations with his compatriots.

Avikson's first article revealed that the military reservists had been conscripted without warning, often in the middle of the night on May 6-7, 1986. The men were said to be disturbed by this method of inducement and asked whether they could not have been taken to Chernobyl somewhat less rashly. Also, some of the men were over 45 years of age, or had families of 3 or more "under age" children, which in theory should have exempted them from service. Many wondered what the Chernobyl accident had to do with Estonians, a disaster that had been caused by operators making blunders in far-off Ukraine. Avikson disagreed with this attitude, commenting that the consequences of the accident went well beyond the capabilities of the Ukrainian SSR. "In such a situation, would Lithuania and Estonia manage alone?" he asked (suggesting, of course, that the conscription was carried out simultaneously in Estonia and Lithuania).[59]

By the time that Avikson arrived on the scene, some of the men had been sent to Khoiniki hospital (on the Belorussian side of the 30-kilometer zone). Their health situation had been accentuated by the sweltering conditions in the day combined with the coolness of the night, and by working in areas of high radiation. He noted that several Estonians had already left Chernobyl and that one group of ten workers was preparing to leave. The over-45 group and those workers with three small children were now being sent home, thus confirming that by military regulations, they should not have been called up for duty in the first place.

The men began their working day, stated Avikson, at 6am. The morning routine involved a meal, travel to the area of work, and then decontaminating various areas. They returned to camp at 8pm. Like their Ukrainian colleagues, they were living in tents which could often be cold and damp, and sleeping on wooden bunks. Their work routine was repeated daily and monthly, with only the 2 rest days each month.

They were working, he wrote "like squirrels inside a wheel." An "odd story" had circulated that the cleanup workers had nowhere to bathe properly after the day's work, that the men were obliged to use nearby streams and puddles and were getting covered with a thick layer of dirt and radioactive dust. However, Avikson, wrote, the washing facilities, all things considered, were quite reasonable.[60] He omitted to add what has already been described above, which is that the facilities were usually a long walk from the area of work.

The work of the Estonian group involved washing down village houses and trees, clearing the topsoil of contamination and loading it onto vehicles, which took it "to the burial places for radioactive substances." He reported that he did not know where these places were because it was a security matter. Elsewhere, it has been reported that the burial site was at the No. 5 Chernobyl reactor. The displaced soil was replaced with soil brought in from clean areas. For the most part, the Estonians worked in areas that were not immediately adjacent to the No. 4 reactor. Some workers, nonetheless did volunteer for work in Prypyat, ostensibly because by working in a zone of higher radiation, they would be permitted to return home more quickly.

A number of problems had arisen, reported Avikson. For example, how water resources could be protected from pollution, and what to do with ramshackle houses that "hardly hold together." The latter could either be decontaminated or simply razed to the ground. It was also implied that there was more topsoil and uprooted bushes than could be dealt with by the containers available. What was to be done with them? In brief, the Estonian situation, initially, was very difficult, but perhaps no worse than that of the other cleanup workers. Like the others, the Estonians had been told that they would be able to return home within a 30-day period, with permission of the Raion Executive Committee and the War Commissariat, but this had not happened.

Earlier, wrote Avikson, "many high officials" who talked to the men had promised them that after two months at the maximum (that is, on July 6), they would be back home. The promise had led the men to believe that the harder they worked, the sooner their return would come about. Some had resorted to the practice of leaving their dosimeters in dusty, contaminated areas for two days to raise their "roentgen readings." Then, in late June, their work period had unexpectedly been extended to six months:

....at first spades were moved furiously, as if every spadeful would dig the men closer to home. And then on a June day, this news....Everyone of course understands about fulfilment of duty...but this was a cold shower nevertheless....Let us say frankly that this caused in the beginning indignation, soreness of heart and anxiety.[61]

The men demanded an explanation of the extension. According to Ilves' account, based on his translation of Avikson, 200-300 men were subsequently involved in a scuffle with the authorities. A work stoppage occurred, which evidently lasted until mid-July, or at least two weeks if one assumes that the men were informed of the decision on the last day of June 1986. The strike and its consequences have been the subject of a furious debate between the Soviet authorities and some Western circles. In the first place, Estonian sources in Stockholm have alleged that the result of the two-week strike was the execution of 12 Estonian cleanup workers. It was also alleged that a Gunnar Hagelberg had been conscripted for duty at Chernobyl from his home city of Tallinn, and had been ordered to guard the reservists, armed with a gun, and to shoot anyone who refused to work. The story was repeated by a reputable newspaper, the Swiss *Neue Zuercher Zeitung*.[62]

Other Western sources also carried fairly detailed accounts about the affair, such as the Dutch newspaper, *NRC Handelsblad*, which maintained that those who refused to go to Chernobyl could be sentenced in accordance with Article 80 of the Penal Code of Law to 5 years of imprisonment. A striker, it maintained, could be sentenced to death and immediately executed.[63] In the summer of 1987, a deputy of the West German Bundestag, who is also a member of the Green Party, reportedly provided materials to the Spanish newspaper, *Cambio-16*, which was based on information he was said to have received from the "Trust Group."[64] In August 1987, a Soviet source provided a translation of excerpts from the Spanish article (as will become evident, it was necessary to provide such a translation in order, in the Soviet view, to prove that the allegations made in the article were untrue):

The government of Moscow passes over in silence the people working at Chernobyl. Instead, information is given out about decorated heroes. Newspapers write about the "patriotism" of those who "sacrificing themselves, entered into the struggle with the nuclear giant." However, the myth about the courageous strugglers voluntarily meeting their deaths to rescue the nation has been shown to be an out-and-out lie. Men and women from all the territory of the Soviet Union were deported to Chernobyl to carry out this work. Tatyana Nazarovna is a medical sister from Kiev, who tended those ill with radiation sickness. "This began in the first days of May," she states, "construction workers, truck drivers, engineers, doctors, by military order were sent into the 30-kilometer zone." Her fiance, Fedya Tikhon, was also exposed. After four weeks, he was released into freedom, i.e., he was already unable to work. He was sick. Only those who are mortally ill are released from the zone earlier than the established six-month term.

Leningrad is located 1,200 kilometers from Kiev. The doctor of a clinic of the First Medical Institute of Leningrad, Sergei Lornov, says: "Many of my colleagues received from the Ministry of Defense an order to go to Chernobyl. One of them returned there after just a month in a lead coffin. He was finished, I measured the radiation dose he had received and knew that he had exceeded the permissible dose by over 100 times."[65]

The article later focused more specifically on the Estonian affair:

Tallinn is located 1,000 kilometers north of Chernobyl. Before the catastrophe Gunnar Hagelberg lived there. Several days after the accident, when Gunnar crossed the street, he was seized by soldiers, and together with hundreds of his compatriots, sent to Chernobyl. They did not even give him the chance to let his family know. There, he was ordered to watch workers in the zone, and in the case of refusal to work, to shoot them. Not knowing anything [about the radiation levels], as he carried out the order, Hagelberg died at Chernobyl. In the 30-kilometer zone operate the laws of the military period. Yurii Medvedev, a member of the Academy of Sciences from Moscow, states: "The military situation at Chernobyl signifies that the fate of the people in the zone, the question whether people live or die, is decided by military tribunal. If people begin to go on strike, they risk being shot on the spot."

The authors of the Soviet journal, *Sovetskaya kultura*, were incensed by the article in the Spanish newspaper, but decided to investigate the comments made. In Kiev, they reported, they searched for Tatyana Nazarovna and her fiance, and discovered that no such people had been issued with passports in Kiev. Next, they began a "raid" on the

Leningrad doctor, Sergei Lornov, and were informed by telephone that no Lornov was listed in the register of assistants at the clinic of the First Medical Institute of Leningrad. A similar story came from Estonia's Ministry of Internal Affairs where, it transpired, there were several families with the name 'Hagelberg,' but none had a family member called Gunnar.[66]

The Moscow Trust members, who were also cited in the account, did exist, but had never been at Chernobyl. The Soviet journal provided quotations from several scientists in the zone, including IAEA Director-General, Hans Blix, and strongly attacked the Green Party. There is no doubt that the *Cambio-16* account was carelessly written, and used false names for its eyewitnesses, probably to protect their identity. It was written in an accusatory and inflammatory style, clearly intended to inspire feelings against the Soviet authorities. Yet, whatever its motives, it did raise some authentic questions. Moreover, these questions were not answered by *Sovetskaya kultura*, which opted instead to discredit the sources, even though the accounts may also have been based partly on Avikson's articles. It was a curious counter-attack in that it did not actually deny anything in the account (perhaps the editors of the Soviet journal felt that it was beneath their dignity to respond to such charges, given the way in which the Spanish article was written).

The report on Chernobyl's cleanup that appeared in the Dutch *Handelsblad* newspaper was sent to Moscow by members of the Dutch Peace Movement. There, it found its way into the hands of Andrei Pralnikov of *Moscow News*, certainly one of the more outspoken and frank journalists working in the Soviet capital. Pralnikov met with Aleksandr Usanov, the Deputy Minister of Medium Machine Building, who had played an important role in the construction of the sarcophagus. Usanov angrily rejected the accusations that several Estonians had been executed for refusing to work—although neither he nor Pralnikov denied that the men had put down their tools. However, Pralnikov found it hard to believe that workers were reluctant to go to Chernobyl or were being sent there by force. He maintained that his colleagues in Moscow who sorted out the editorial mail coming to his newspaper had come across thousands of requests from all over the Soviet Union stating: "Help us go to Chernobyl. We want to participate in eliminating the accident."[67]

Similarly, Gennadii Lykov, the head of the construction department concerned with the erection of the sarcophagus, had noted in September 1986 that almost all those who had received permission from medical specialists remained in Chernobyl for an additional time period in order to complete the work they had started: building the concrete shell over the reactor. In other words, the cleanup workers were even anxious to remain in the zone beyond their official timespan. Again, however, the approach avoids many of the questions raised. Extending one's voluntary work period is not the same as being conscripted for a certain period and then having it trebled or raised by six times over the initial level. Moreover, the fact that there were numerous volunteers to go to Chernobyl in no way negates the possibility that those who were there were anxious to leave. It is not uncommon for expectations about a job to be dashed quickly, particularly when the work was as harsh as at Chernobyl.

But let us raise another question. Were there any other reports from the Soviet Estonian press that indicated that conditions in the zone were a severe trial for those involved? The answer is that even those workers who volunteered found conditions close to unbearable. One such volunteer was Yurii Baydyuk, who from his name may have been a Ukrainian living in Estonia. He was part of a volunteer group that was formed in Estonia in July 1986. Upon arrival near Martynovichi, on the flood plain of the Uzh River, the group had built itself a small house with a stove for heat and to dry clothes. On the third day, the group began work to build a dam that was 6 kilometers long, 15 meters wide and 5 meters high and to insert filters in it (presumably for the 1987 spring floods):

> The sun was hot, sweat flooded the eyes, there were mosquitoes and gadflies, people were breathing in respirators and had headaches. We had to battle mud and swamp water; the mud kept devouring the fill and we had to keep pouring it in in order to deposit any material. At nights, after our shift, we had to unload flatcars full of quarry stone. In the morning it was back to the dam. It was like this every day, without holidays, for three months. I couldn't get used to it. It wasn't the exhausting work, but the devastation around me, the oppressive silence.....[68]

When the task was completed, however, Baydyuk and his colleagues had one final task, which was to drop pieces of graphite from the roof of the nuclear plant's fourth unit onto the reactor. The radiation levels were so high that they could only work for 40 seconds at a time. We have noted this same task above. What is remarkable in this instance is that men were given such a task after having been in the special zone for 3 months. Earlier accounts had suggested that the roof work was the sum total of that of any worker who agreed to do it. In the end, Baydyuk comments, his fatigue was such that he and his colleagues were kept going only by their consciences.

Another account of the life of the reservist appeared in an Estonian raion newspaper, *Tartu Edasi*, which interviewed a 45-year old man, by trade a tractor driver, who was sent to Chernobyl on the orders of the War Commissariat, on May 7, 1986. According to Ilves, the article gave the impression of being heavily censored because several passages no longer made sense as read. Nevertheless, a portrait emerges of a man who was discontented with his lot who was encountering very high radiation levels even though working some distance from the No. 4 reactor. The man, August Lepik, and his colleagues decontaminated houses and settlements, washing away radioactive dust and delving into places that could not be reached by machines, with shovels.[69]

Again, the arduousness of the work is stressed. Not only did Lepik work 12-hour days without a day off for almost two months, but he could not sleep at night because of "constant noise." Even at night he stated, there were helicopters flying over to test radiation levels and armored cars carrying out their work. Lepik was released before his colleagues, either because he became ill or, as seems plausible, because of his age. The article is notable for supplying another story that verified many of the points made by Avikson. Together with Baydyuk's account, they undermine the assertion that the cleanup workers were a satisfied crew. Taken together, the cleanup workers seem to have been, at the least, discontented, and at times even mutinous. Their voices, however, were not heard very often. Even in the period of glasnost, it was left to relatively obscure sources to bring their stories to light. One suspects that the vast majority of stories about the cleanup operation have never been told.

In the summer of 1987, as a sort of postscript to the Estonian affair, there were some new developments. An Estonian samizdat document appeared in the West which stated that in February 1987, Toomas Leito, the head of the Propaganda and Agitation Department of the Central Committee of the Communist Party of Estonia, had lost his position for permitting Avikson's articles to appear in *Noorte haal*. The account maintained that the chief complaint against Leito was his "incorrect interpretation" of the articles, which had caused something of a sensation in the Western press. As Ilves points out, Avikson's revelations appeared almost at the same time as the IAEA meeting in Vienna, which for a variety of reasons was a major international success for the Soviet Union. The articles were a veritable "fifth column" in the rear of a propaganda coup. That Leito lost his job was confirmed by the Plenum of the CC of the Estonian Communist Party in March 1987, at which time he received a lower ranking post with another newspaper. The samizdat document noted that Avikson himself was not punished and reappeared as a columnist in *Noorte haal* in late-September 1986.[70]

In July 1987, a broadcast on Radio Stockholm announced that Karl Kimmel, the Estonian SSR State Prosecutor had confirmed that the head of the Estonian military establishment, Major-General Roomet Kiudmaa, was under investigation for bribery. Kimmel had reportedly informed the Swedish radio station by telephone that a preliminary investigation was under way, but that "final charges" had not yet been laid.[71] A Western news agency declared that according to "private sources in Tallinn," Kiudmaa was accused of selling deferments from duty in Afghanistan for 1,000 rubles, and from duty at Chernobyl for 500 rubles.[72] The bribery had reportedly come to the attention of the Estonian Minister of the Interior, Major-General Marko Tibar, who had been "forced to resign" and leave the Communist Party because he had declined to intervene in the scandal.[73]

Once again, the Soviets responded to the charges. Radio Moscow maintained that the reports of Kiudmaa's arrest had been "deliberate fabrications." It admitted, however, that the State prosecutors had been investigating allegations that Kiudmaa had accepted bribes, although it did not state what the bribes might have been taken for. Radio Moscow also rejected Swedish reports that Tibar had resigned for failing to stop the bribe taking. Marko, it declared, was on vacation, and he had no intention of resigning.[74] There, the saga ended. Its significance lies not in the investigation itself, but in the plausibility of the charges. It indicates that Chernobyl was, in the eyes of some recruits at least, scarcely less unattractive than Afghanistan and that people would have been willing to pay between one-quarter and one-half of an average Soviet monthly salary (possibly higher for the reservist) in order to avoid cleanup work.

The Estonian situation remains something of an enigma, like many other aspects of the Chernobyl disaster. But it is only the conclusion to the events that is in doubt, not their occurrence. The official Soviet history of Chernobyl will not include the events described by Avikson and others, yet their story, despite subsequent embellishments by those who would wish to see the Soviets discredited for personal reasons, should not be forgotten. It is not a pleasant one, just as the cleanup work itself, which placed the lives of all the personnel involved in danger, was not a pleasant episode. That thousands were conscripted for work in the zone by force is not in doubt. Nor are there doubts about these recruits incurring high levels of radiation during the course of their work.

The only doubt is the major question about the entire operation: was it necessary? The sarcophagus aside, was it really essential to throw forces into a contaminated area so soon—10 days—after a major disaster? Could not the authorities have left the area alone to let the natural processes of elemental decay take place for, let us say, 2-3 years, before attempting a cleanup operation? Moreover, when the operation did take place, surely forces could have been used more widely, without forcing reservists to remain in the special zone for six months at a time?

One cannot expect an answer to these questions, since thus far the Soviet authorities have not accepted their validity. In their view, it was essential to mount an immediate cleanup campaign for three major reasons, which were as follows (although not necessarily in order of importance):

1. As has been stated on more than one occasion, the present generation has a duty to descendants to minimize the disaster's impact on
   the natural environment. No empty wasteland can be permitted to
   emerge. If the zone had been left alone for any length of time, then
   in effect it would have become a dead zone because its former
   residents would have become employed or occupied elsewhere in the
   republic or in the country. Thus, it may have been a case of "now
   or never" for the cleanup, since former residents could not have been
   expected to reside indefinitely in temporary accommodations.

2. The psychological aspect of the disaster was always uppermost in
   the minds of the Soviet authorities. If it could be shown that Chernobyl had been a triumph rather than a major setback, that Soviet
   citizens, united and determined, had "beaten back" the effects of the
   atom, then this could be regarded as breaking a psychological barrier. It would forestall panic among the population, fear of radiation, and above all, fear of the atom in an uncontrollable state.

3. The authorities did not want to abandon the Chernobyl nuclear
   power plant, partly for economic reasons and partly for military
   reasons. A deadline of October 1, 1986 had been established for the
   return of the station to service, and Usanov, the Deputy Minister
   of Medium Machine Building was on hand to organize some of the
   work on the sarcophagus, to ensure that the timetables were met.
   In the event, the station came back on-stream as scheduled, even
   though the conventions that followed the IAEA meeting in Vienna
   were still continuing, and even though the sarcophagus had not yet

been completed. This latter circumstance was probably one of the most astonishing developments of Chernobyl, but coming after the Soviets' public relations triumph in Vienna, it did not have the impact it might have had otherwise. Yet it is a fact that the Chernobyl-1 unit was kept in operation a full day after the explosion ripped off the roof of Chernobyl-4, and it was back in operation two months before Chernobyl-4 had been properly sealed. There could be no clearer indication of the vital importance of the Chernobyl nuclear power plant to the Soviet authorities in Moscow.

In conclusion, the special zone was a unique place, and it is possible, as one source stated, that within its confines, "nobody smiled." Still, the cleanup crews may have smiled ruefully had they read a story in *Sovetskaya Estoniya* about the Baltic ensemble "X-ray," which had been formed at a settlement built in Zhytomyr Oblast and performed its first concert for shift workers in Prypyat. Why the strange name?, asked the writer. "If the men were really in constant danger, then they probably would not be joking about it."[75]

6 Restoration and Reconstruction

The 30-kilometer zone around the Chernobyl nuclear power plant included virtually all of the Chernobyl Raion of Kiev Oblast. The raion, which stretches northward to the border with the Belorussian SSR, has a total area of 2,000 square kilometers, and a population of approximately 90,000. Before the disaster, approximately half of the population, which lived in 70 different population points, resided in the city of Prypyat. The latter city is located some 3 kilometers north of the Chernobyl plant. After the accident, therefore, the world's attention was drawn to a relatively remote area with a small population, making up only about 5% of the total population of Kiev Oblast (excluding the major city of Kiev). The Chernobyl Raion is dissected from northwest to southeast by the Prypyat River, which flows by Kopachi and Leliv before reaching Chernobyl city, and then passes by Ladyzhychi and into the massive Kiev Reservoir.

This area has proved to be a major headache for the Soviet authorities, for several reasons. First, it threatened to become a "dead zone," a permanent monument to the nuclear accident. Second, its residents had been uprooted and, for the most part, were anxious to return to their homes. Third, as noted, the Chernobyl nuclear power plant dominated the industry of the raion. It "ate up" workers and plant operatives. Thus at 66% of its proposed capacity in April 1986, it required the services of operatives, building workers and their families to the extent that the population of Prypyat rose to 45,000. One of the USSR's new "nuclear plant" cities, Prypyat, like the others—such as Enerhodar in Zaporizhzhya and Netishyn near the Khmelnytsky station—was expected to continue its rapid growth. Generally, the building of city amenities followed rather than preceded the callup of builders to work on reactor units. Nonetheless, by 1986, the authorities were proud of Prypyat, with its tree-lined streets, kindergartens, parking areas, theater and other amenities. It had been in existence for only fifteen years and the authorities were reluctant to abandon it.

Some 22 kilometers to the south is Chernobyl itself, with a relatively stable population of about 11,000 in April 1986. Chernobyl's past history as a small port played a major part in Soviet thinking in the post-accident situation. Only a limited amount of materials could be delivered to the area by roads, particularly in the spring when floods made them muddy and impassable. But one could reach Chernobyl from Kiev by boat. Not only were supplies brought into the zone in this manner, but, as will be shown, boats were used to accommodate both the cleanup crews and workers at the nuclear power plant for several months over the summer of 1986.

In May 1986, however, the first concern of the authorities was where to take the evacuees and, since they could not all be returned to the zone in the short-term, where to rehouse them. The raions immediately to the south of Chernobyl were designated for this purpose: Ivankiv, Borodyanka and Makariv in particular. The logic to this was that the agricultural population would remain in familiar territory, and the move could be made with the minimum of fuss. In the longterm, a return to the former home would not provide a major problem, while in the meantime, when conditions improved, the evacuees could easily go home and collect their belongings. Outside the Kiev Oblast, those evacuated from the southern parts of Gomel Oblast in Belorussia were, for the most part, being moved to the northern part of the oblast, which had been suffering from a shortage of labor. In addition some dwellings were to be built and empty houses made use of in Zhytomyr Oblast, on the western border of Kiev Oblast.[1]

New Villages

In early September 1986, it was reported that Account No. 904, which had been opened by the Soviet government to assist the victims of the Chernobyl disaster, had raised 490 million rubles. Of this amount, 100 million rubles were being used to build about 7,000 new homes for evacuees in Kiev Oblast; and 100 million rubles were put toward financing 3,960 new homes in Gomel Oblast. In the latter area, it was said that the total compensation paid to each family for the loss of their homes, their property and their livestock amounted to about 14,000 rubles per family, although as many of the families who were interviewed pointed out, this amount of money did not last long, nor was it always forthcoming.[2]

The first new village to be constructed was revealed with fanfare in August 1986. It was located 4 kilometers south of Makariv, the center of the Makariv Raion, very close to the tiny village of Lyudvynivka, and about 130 kilometers south of Chernobyl. The huge Gorky collective farm, which was mentioned above (Chapter 4), was located nearby. On August 2, hundreds of people stood on the square of the new settlement, which had been built by construction workers from Ternopil Oblast in western Ukraine. It had been built for evacuees from Chapayevka and Horodchany, located in the northeast of the Chernobyl Raion, some 20 kilometers from the nuclear plant. It was reported that as early as May 1986, it was clear that these evacuees would not be able to return to their homes.[3]

In less than 2 months, a new village of 150 brick homes had been built. The schedule for construction had not been quite so rigorous. The original timetable had called for the completion of construction by September 15:

Above the headquarters hung a slogan: "Ternopilites! With shock labor we will help the Chornobylites. Let us complete good-quality homes by 15 September.[4]

The date on the banner had then been crossed out four times; at first, to August 24, then to August 10, and finally to August 1. The local collective farm chairman (also noted above), Mykola Tyapko, had provided various assistance, from materials to men. After a downpour, for example, the soil walls of the trenches containing water pipes had collapsed, and he had provided workers to rebuild them. At the opening ceremony, the Kiev Oblast First Party Secretary, Hryhorii Revenko, cited the role of Tyapko in helping to build this, the first of 52 new settlements being erected in the oblast. The settlement was named Ternopilske [Ternopolskoye in Russian, the name that was used by Western sources], as a "token of thanks" to the builders.

At the ceremony, an elaborate ritual was played out, ostensibly to show the new residents that this was now their permanent home and not something makeshift. It was as though it was psychologically necessary to demonstrate to the evacuees the permanence of their situation in view of the fact that on most occasions, the Soviet authorities had emphasized that the population transference was a temporary affair. Thus a native Ukrainian song was played and a red ribbon cut to a round of applause, and then the Chernobyl area residents walked onto the square, the men in cloth caps and jackets and the women in colorful native costume, "embarrassed by the honor." On the porch of each home, two young girls in Ukrainian national dress, wearing wreaths of flowers on their heads, gave the new proprietors of each home the traditional welcome of salt and bread, along with the keys to the new home, and wished them health and happiness.

On the previous day, Tyapko and the chairperson of the collective farm whose residents had been transferred to Ternopilske had visited the new homes. It was reported that each possessed a farmstead with a barn for cattle and fowls, with a cellar made of ferrous concrete, a stack of firewood, a bear spear for drying pitchers and pots, a doghouse and even a tree stump on which to split wood (1,501 tree stumps had been brought into the new settlement). A mailbox had been inserted in the wicketgate of the fence. Each house reportedly had gas in cylinders, electric lighting, beds with linen, towels and mirrors. The kitchens were adorned with a table, 3 stools, various utensils, pails, jars of salted food, meat, stew, sacks of buckwheat and rice containing about 30 kilograms of each, and there were potatoes in the cellar. In short, the report noted, the new homes, which cost about 27,000 rubles each (U.S.$47,250), had "everything that a villager needs!"

Over the porches of the new homes horseshoes had been hung for good luck. Concerning superstition, frequent references are made in Soviet accounts to storks. Ternopilske was no exception. On the porch of his own new house, which was said to be similar to the one described above, the former collective farm chairman from the Chernobyl Raion, Tyapko, saw a stork in the sky. The builders had installed a pillar in the village for the bird to land on so that it might settle in the new village, guard over its future happiness and prevent misfortune. The stork evidently circled the pillar and flew away, "but will return next year with a mate to nest."

By mid-August 1986, about 2,500 new houses had been built for evacuees from the 30-kilometer zone (on both sides of the Ukraine-Belorussia border), while a further 4,750 had to be built by October 1. In addition, 8,000 apartments were made available in the cities of

Kiev and Chernihiv for families from Prypyat.[5] The progress on the work was examined not only by the Government Commission, but also during high-level visits from members of the CC CPSU Politburo. In August, for example, Soviet Premier, Nikolai Ryzhkov, and KGB Chairman, Viktor Chebrikov, visited some of the new homes, and reiterated that all families should have received them by October 1.[6] Despite the public relations success of Ternopilske, matters did not always run smoothly in the allocation of the new homes.

For example, in late October, a Soviet report stated that some of the evacuees had not yet received the homes they wanted, while in other instances, people had tried to use the disaster as a means "to wangle" a new home or apartment to which they were not entitled. One retired couple had been unable to find accommodation in the village in which their married daughter had been lodged. In Prypyat, the two families had been neighbors. As a result, the older couple had refused their allocated quarters in a nearby town, and had also rejected a suggestion that they move in with their daughter. It was also stated that there were people who had been dismissed from their jobs before the accident, but who had alleged that they were still entitled to one of the new homes on the grounds that they had been fired unjustly and should have been working in the special zone at the time of the accident.[7]

Similarly, the head of the Chernobyl area's agricultural energy program had evidently used the disaster to award himself a 2-room apartment to which he was not entitled, and he was therefore removed from the party. On the other hand, the authorities were having difficulty in persuading some "professionals" to move into the new houses. The new settlements required doctors and teachers, for example, and 600 houses had been reserved for people in these professions, but remained empty in late October. Moreover, one huge apartment building had been set aside for the evacuees, those who were originally registered for the rooms having given them up for the victims of the disaster. It remained vacant, however, and the planning authorities were said to have miscalculated in appropriating the building for evacuees.

By mid-December 1986, about 12,000 new homes had been built for the evacuees.[8] Of this number, over 4,000 were on the Belorussian side of the border. However, the process of moving evacuees into new settlements was much slower north of the Ukrainian border. In fact, by early January 1987, only 102 of the new homes were actually inhabited, or 2.5%. One reason for the delay may have been in erecting facilities for the new settlements. Concerning the settlement being built near the village Kalinkovich, for example, a situation of "total chaos" was said to exist as dump trucks, diggers and other machines were lining up for fuel, all heavily laden with concrete, bricks and other materials. The Chairman of the Belorussian Government Commission dealing with the accident's consequences, the First Deputy Chairman of the Council of Ministers of the Belorussian SSR, Vladimir Evtukh, descibed a "very tense" situation, and said that the needs of senior residents in particular needed to be attended to.[9]

Some idea of the numbers involved in the rehousing process was gleaned from statements made by I.S. Plyushch, Chairman of the Kiev Oblast Executive Committee, in April 1987, although his figures were somewhat puzzling. In Kiev Oblast, the 52 new villages had accommodated 27,000 of the reported 92,000 evacuees by the late Fall of 1986. Approximately 22,500 other evacuees were living in apartments in Kiev—Plyushch did not mention Chernihiv, but presumably the figure would have included the smaller number who were living in that city too. Also, for former residents of Chernobyl and Prypyat, some 13,500 apartments had been assigned in both new buildings, and in raions and cities of Kiev Oblast. In the summer of 1987, it was planned to construct a further 3,000 farmstead-type homes and to provide another 1,500 apartments for the permanent settlement of evacuees. In the Ukrainian SSR as a whole, it was stated that almost 34,000 *families* had been resettled.[10]

What can be deduced from these figures? The first response is that if they are accurate, then the number of those evacuated must have been considerably higher than officially stated (as we suggested in Chapter 1). The average family size appears to have been about 3.3 persons. Therefore, approximately 112,000 persons had been rehoused by the spring of 1987. Yet, as Plyushch stated, in several areas conditions were still cramped because two families were living in tiny homes, not having been rehoused. The new homes and apartments to be made available by the summer would account for a further 14,850 persons being rehoused, giving a total of 126,850 evacuees from the Ukrainian side of the 30-kilometer zone alone.

But not all those evacuated remained outside the zone. The staff of the Chernobyl nuclear power plant, for example, was reported to have 90% of its pre-accident personnel back in place by August 1987.[11] Although their families would not have accompanied them, this raises still higher the overall total originally removed from the zone, which even from the most conservative estimates based on the above figures, would appear to be over 162,000 (including Belorussia).

Some of the new homes were displayed to foreign journalists in the summer of 1987, a period of notable openness on the part of the Soviet authorities. For example, the settlement at Nedro (Baryshiv Raion) had developed from the existing village, and was something of a model facility, with wide streets, public baths, and a trade-cultural center.[12] Some 40 foreign journalists were permitted to visit the village of Zdvyzhivka, in Borodyanka Raion, which accommodated evacuees transferred from Zalissya, 3 kilometers south of the city of Chernobyl. There were 250 homes here for the new settlers and the journalists were able to interview pensioners Olga Antonovna and Kuzma Ilchenko, who declared themselves happy with their new home and with their compensation money.[13]

A Western reporter, Christopher Walker of *The Times*, visited the new village of Nebrat, which houses over 1,600 evacuees, and wrote that it was "typical of the soulless new villages" that had sprung up on former farmland. In his view, the amount of capital invested in the new homes demonstrated the permanence of the sojourn away from the 30-kilometer zone.[14] Correspondents from the Western press also visited the new village of Tavrya, which accommodated residents of a former collective farm on the border of the 30-kilometer zone. Here, the houses were said to be identical, pink in color, and to "line 3 dusty streets." Work on the new village had reportedly continued for 24 hours a day in order that it might be completed by September, when children were scheduled to return from summer camps on the Black Sea and other resorts.[15]

When evacuees were rehoused outside Kiev Oblast, life appears to have been more difficult. Away from the glare of the cameras of Soviet and foreign correspondents, it was somewhat harder to draw attention to the problems encountered. One story that did come to light concerned Chernobyl evacuees who had been moved eastward to the mining town of Rovenky, in Voroshilovhrad Oblast, where a section of an apartment building had been made available for them. Built by the Sverdlov Home Construction combine, the quality of the new apartments had been evaluated as "adequate." However, in the part of the building that was to be inhabited by local residents, it was reported that being "well informed," they had stocked up on all those necessary items that they knew would be missing when they moved into their new homes. Chernobyl-area residents, who had left most of their belongings behind, were less fortunate.[16]

One evacuee referred to the "formalism and indifference" with which they had been met upon arrival in Rovenky. The builders had left "many things incomplete" in the new apartments, which led to a "ruined mood" instead of joy. Doors, windows and the heating unit had not been painted. Linoleum had been placed directly on top of builders' garbage. Plinths had not been fastened. The heating system was functioning, but the temperature in the rooms never exceeded 10 degrees celsius (50 degrees fahrenheit). There was no hot water available. Several commissions had been sent to inspect the new apartments, and the Deputy Chairman of Voroshilovhrad Oblast Executive Committee, L. Daineko, had also paid a visit. The Chairman of the City Executive Committee, V. Ponomaryov, had evidently dismissed the complaints of one distressed evacuee, who had two small children and a husband still hospitalized, as emanating from her complex character. A Kiev newspaper, rejecting the dismissal, stated that:

> In connection with this, we would like to state that neither the superfluous emotions of people whose fates have been afflicted, nor caring letters to the editor, nor telegrams would have existed in the above instance if one of the authoritative comrades of this city executive committee had looked personally at the apartments given to families that suffered from the accident at Chornobyl atomic energy station. Then the trouble today would not be dealt with by formalism, but by substance, and the leadership of industrial enterprises concerning the problems of these families would also be substantive. At the least, these builders would have organized a subbotnik to rectify the situation.[17]

Despite these problems, and the above dilemmas are not untypical of Soviet apartments generally, the majority of evacuees had been reaccommodated very rapidly over the summer of 1986. The cost of the new housing alone amounted to over 324 million rubles, which had to be added to the various amenities required in each new settlement, and the amount of personal compensation paid to each individual family. The houses were built rapidly so that families were no longer billeted by the time the frosts arrived. The pace of building in Belorussia, for example, was said to be faster than the estimates could be put together, and houses were built not singly, but five at a time. The evacuees, as demonstrated earlier, were not particularly happy about their situation, but there is no doubt that the authorities, using construction organizations from throughout the Ukrainian and Belorussian republics, had moved ahead with all possible speed. Such a policy was commendable. What was surprising was that although by the summer of 1987 there was apparently sufficient housing available to accommodate all the Chernobyl evacuees, simultaneous policies were also in effect directed at the ultimate revival of the 30-kilometer zone, almost in its entirety.

The two policies were contradictory and suggest that the Soviet government was pursuing more than one option at a time. The zone would be revived. But if it was ascertained that the contamination was too severe to undertake such a policy, then the new cottages would remain the permanent homes of the evacuees. The former residents of Prypyat, including the families of Chernobyl operatives and builders, were for the most part residing in apartments in Kiev or Chernihiv. Their future was decided early in October 1986, when—some have stated against all the odds—the Chernobyl plant was restarted only five months after the major disaster.

Restarting the Chernobyl Nuclear Power Plant

The accident at the No. 4 Chernobyl reactor had not affected the first two units at the station, which continued to operate even while the furious struggle against the fire continued at the damaged unit, and threatened to spread to the roof of the No. 3 reactor. On the following day Chernobyl-1 and Chernobyl-2 were shut down. Thereafter, the Soviet authorities always maintained that the goal was to bring the station back into operation by October 1986, despite the formidable problems entailed. As we have shown, the timetable was dependent upon the sarcophagus being built, but not upon it being completed. It was less dependent upon the deliberations of the IAEA conventions in Vienna, even though the Soviet delegation was participating actively in the discussions about safety of nuclear power plants. On the other hand, the timetable was based upon the period needed to add modifications to the RBMK reactors (Chapter 7).

On August 29, the Ukrainian First Party Secretary, Volodymyr Shcherbytsky, and the Ukrainian Premier, Oleksandr Lyashko (subsequently replaced), visited Chernobyl nuclear power plant and had talks with Gennadii Vedernikov who was at that time Chairman of the Government Commission dealing with the accident's aftermath in the Ukrainian part of the special zone. Among the topics discussed was the renewed exploitation of the first and second reactor units and the progress of decontamination around those reactors. Shcherbytsky demanded a more rapid tempo of work, so that scheduled targets could be met in the agreed time.[18] Three weeks later, Borys Shcherbyna of the USSR Council of Ministers, who had played an active role in the post-accident operations, stated that the Chernobyl-1 unit had been modified and that the crews had been retrained since the accident. He also noted that the Chernobyl-2 unit would also be ready for operation soon.[19]

The Soviet media then waited anxiously for the inevitable announcement. On September 29, the first unit was started up on a test mode before being switched to commercial operation. Along with the announcement, Radio Kiev assured listeners that the burial of Chernobyl-4 was being completed, which was hardly accurate. Two days later, at 4.48pm, Chernobyl-1 was officially restarted, on the exact day scheduled, and power climbed to 40 megawatts.[20] On October 1, Radio Moscow interviewed Yurii Semenov, a Deputy Chairman of the Government Commission, who announced that after major work on the repair and rehabilitation on unit 1, its No. 1 turbogenerator was being brought back into service. By October 2, the power had been raised to 300 megawatts.

Izvestiya had sent reporters to the site of Chernobyl-1 on October 2, 1986. They were informed by the shop chief Shadrin that the shift-work schedule of the operatives had been amended. Instead of the 2-week shifts followed by 2-weeks' rest, the workers now worked for 5 days, followed by 6 rest days in order to keep in touch with events. There were still serious problems regarding housing for duty personnel (see below). Just before the startup of Chernobyl-1, changes had been made to the reactor's loading system and the safety interlock system. One of the interlock safety units had been inadvertently triggered and the reactor had shut itself down automatically again. The shutdown process had taken 10 minutes, but the repair operation took two days, which, said the speaker, gives one "an idea of how carefully every step is being performed."[21] Although not resulting in a dangerous situation, the sudden shutdown was also an ominous reminder of what had occurred five months earlier.

As with the completion of work on the first new village settlement for evacuees, a rally was held to celebrate the revival of the Chernobyl nuclear giant. By October 2, at 9am, the second turbine at Chernobyl-1 had been started, and the unit was now "fully on-line." Usanov turned up at the rally to issue awards on behalf of the Government Commission to "outstanding workers." But the rally was also used as an appeal to those involved in the cleanup around the Chernobyl units to finish the work by their official deadlines as unit 1 approached full power. Chief Engineer of the Chernobyl plant, N.A. Shteynberg, announced that unit 2 would also be brought back on-line as soon as repair work on it had been completed.[22] Also on October 2, the situation at Chernobyl was analyzed in a session of the CC CPSU Politburo, which evidently declared itself satisfied with the measures taken to decontaminate the affected territory from radioactive substances and with the preparatory measures for refiring Chernobyl-1 and Chernobyl-2.[23]

The impact of the event on both the USSR and the Western world was considerable. The U.S. Minister of Energy, James Herrington, stated that he was worried about the restarting of Chernobyl so soon after the major accident. By contrast, the Three Mile Island nuclear power plant, at which an accident of less significance had occurred in 1979, was still shut down. Soviet spokespersons, such as Yurii Semenov, had no time for such concerns:

> Well let him [Herrington] worry about the work of his own nuclear power plants. Chernobyl nuclear power plant is operational. Today's introduction of the No. 1 block into the network should make various soothsayers bite their tongues.[24]

A more balanced, but nonetheless proud account of the events preceding the restoration of unit 1 was provided by Erik Pozdyshev, the then Director (from May 1986 to February 1987, and currently a worker in the Ministry of Atomic Power Engineering of the USSR in Moscow):

> The startup of unit 1 was a kind of psychological landmark. The station is coming back to life. And everything must function flawlessly. We have tried to make sure that when nuclear plant personnel—workers, engineers and scientists who worked here before the accident—arrive here today, they can feel immediately at home.[25]

Before the unit came on-line, Pozdyshev added, proper work conditions had to be restored. The topsoil had to be removed at a depth of 30 centimeters, and the entire nuclear plant site had to be covered with concrete slabs and sealed because it was covered with contaminated dust. But at the first possible moment, repair work was begun and the additional safety measures were implemented at the first two units. The social facilities for the workers had remained from before, and Pozdyshev had even insisted on the repair of the marble staircase in the lobby. In short, Pozdyshev, the Director, was immensely pleased with the progress made. He maintained on another occasion that analogous work had been carried out at Three Mile Island, but that whereas it had only taken five months to cover the tasks at the Soviet station, it had taken the Americans five years to reach the same stage.[26] The comparison, however, seems unbalanced since the Americans did not have to encounter major external contamination of the surrounding area and adjoining countries. In fact, Pozdyshev's bravado over the brief schedule only raises further questions about the thoroughness of the work carried out, particularly on the units themselves.

During his interview on Radio Moscow, Semenov had stated that Chernobyl-2 would be back on-stream within 10-15 days, with Chernobyl-3 being returned to service by June 1987. The forecast appeared optimistic, but by November 5, a test run on the No. 2 unit was begun. What was meant by the word "test"? According to the representative of the Ministry of Atomic Power Engineering at the unit, Yurii Filimontsev:

> These tests are not a result of excessive caution on our part. They are designed to make sure the unit is completely safe and will be strong enough for normal longterm operations at Chernobyl nuclear power plant.[27]

By November 10, power at unit 2 was up to 650 megawatts, It was run at peak load for about five months and then shut down again for what were described as "precautionary repairs."[28] On another occasion, a Kiev source that is published in English, interviewing plant Director, Mikhail Umanets, referred to this process as "routine maintenance,"[29] although this procedure is usually carried out after at least one year of operation. Perhaps the reactor was still undergoing some modifications. Nevertheless, the targets for 1986 had been reached. The station was back in service with a capacity of 2,000 megawatts, which was of inestimable service to the Ukrainian power grid during the winter period. At the end of 1986, prospects for units 3, 5 and 6 looked good, and in fact Pozdyshev stressed that once unit 3 came into service, then units 5 and 6 would come on-stream in short order:

> The future is quite plain. After the No. 1 unit, we will complete preparations for the startup of No. 2 unit—the country needs the electricity. Next year we will start up the No. 3 unit. And later Nos. 5 and 6 units, whose construction was disrupted by the accident.[30]

A month later, Pozdyshev was equally optimistic. Unit 3, he declared would definitely resume operations in 1987, and unit 5 would also be ready for operation during this year.[31] Yet between the end of 1986 and the spring of 1987, and particularly during the month of April 1987, there were serious discussions among the Soviet nuclear authorities that resulted in a radical reversal of the planned program and in the case of Chernobyl-3, a delay in returning the reactor to full operation.

Looking first at Chernobyl-3, Andronik Petrosyants, without reference to the earlier forecast of June 1987 as the operation date, stated in late April 1987 that there was a considerable amount of decontamination work ahead on the reactor unit and that it would be started up at the end of 1987 following prophylactic measures and "supplementary safety procedures."[32] An even more cautious prognosis was given by M. Rylsky of Novosti Press. He stated that decontamination of the equipment and buildings was continuing on Chernobyl-3. There had been no damage there, but the enforced shutdown and the proximity of the reactor to unit 4 necessitated the prophylactic repair of technology and the replacement of equipment that had been damaged by the accident. He would say only that "it was proposed" to operate the reactor at a nominal capacity by the end of the current year.[33]

Not until August 1987, when Kombinat spokesperson Kovalenko was interviewed by various news agencies was it made clear that one reason for the delay in restarting Chernobyl-3 was that radiation levels in the area were still very high. Even 300 meters from the sarcophagus, he stated, radiation levels were "from 12-14 milliroentgens" (hourly).[34] Immediately after the accident, Soviet sources had declared that the highest levels of radiation in the zone were 15-30 milliroentgens per hour. These figures are now acknowledged to have been erroneous.[35] The real radiation levels were around 1,000 roentgens (1 rem) per hour, and it was this level to which Kovalenko compared the 12-14 figures noted above. But even 12-14 milliroentgens per hour is an increase of 2,800 times over the normal background level, more than 15 months after the disaster. Unit 3, it can be safely surmised, was physically ready to be returned into service, but the radiation conditions were preventing this from happening.

Nevertheless, on December 4, 1987, Chernobyl-3 was restarted. Umanets maintained that it was refired following "serious and complex" decontamination and repair work, in addition to scientific research. It was true that radiation levels there remained higher than normal. This, according to one account, had led to the question whether it would not be better simply to bury unit 3 alongside unit 4, rather than expend such huge costs and manpower for its restoration. Ye.I. Ihnatenko, the General Director of the Kombinat association, responded by declaring that the "sceptics" had not considered various factors of both a technological and economic nature. Thus to preserve the reactor unit indefinitely by shutting down the reactor permanently would require just as much expenditure as its preparation for exploitation. The implication of these remarks is that the costs could be partly recouped from the electricity generated from the "live" reactor, whereas there would be no compensation for a "dead" one. Thus, said Ihnatenko, "One can say that the realists have won a victory."[36] Yet this was hardly the whole issue. Of more significance was the reemployment of plant operatives who would be forced to work in a zone of enhanced radiation background very close to the sarcophagus that was formerly unit 4.

With units 5 and 6, it seems that a variety of factors came into play. In an interview published in April 1987 (but probably consigned to press in March 1987), V. Komarov, the Chief Engineer of the Kombinat association, remarked that the industrial sites of Chernobyl-5 and Chernobyl-6 were being cleaned up, because the energy they would be producing was essential to the country.[37] A week later, he would say only that building operations on the two units were not being carried out, "we are merely considering the possibility." Expert opinions and data were needed.[38] Then on April 25, 1987, Petrosyants announced on Radio Moscow that a decision on whether to complete and bring into service the two units had been delayed until the 13th Five-Year Plan (1991-95). The two reactors remained in the Soviet nuclear energy program, which means that when speaking of future capacity increases, Chernobyl's fifth and sixth units would always be included in the total figures, even though their future was uncertain.

Late in March 1987, a major conference was held in Kiev to discuss the future of Chernobyl-5 and Chernobyl-6 that was attended by 60 scientists.

Evidently, "specialists" had gone ahead before the accident with plans to build fifth and sixth units, despite accusations of "gigantomania" from certain quarters in the Soviet Union. Chernobyl-5 had been scheduled to start operating in the Fall of 1986, but building work was halted after the Chernobyl-4 explosion.

Public discussion of the plans to go ahead with the new units—the "third stage" of construction of the Chernobyl plant, since its reactors are built in twin sets—was organized by Ukrainian departments and sections concerned with nuclear energy. In the past, as a Soviet source noted, similar questions were not subjected to public approval because the opinions of the authors of such plans had been considered "indisputable guarantees" of their correctness. In fact the meeting in Kiev was attributed to the expansion of glasnost in the Soviet Union. The "democratic beginnings" in Soviet society and the "bitter lessons" of Chernobyl had brought about the decision to subject plans to a wider array of opinions—and what is more, it appears that for the first time in Ukraine's nuclear power history, Ukrainian scientists were permitted a significant input into the decision on the largest of their nuclear power plants.[39]

At first, the chief engineers of the expansion, the Directors of Kharkiv and Moscow sections of the "Atomic Energy Planning" (*Atomenergoproekt*) Institute presented the case for going ahead with the new units. It appeared that the third stage of development at Chernobyl was to be very similar to that of the second stage (Chernobyl-3 and 4), with the addition of several extra safety features, such as additional diesel generators and extra protection for the reactor, and, perhaps most important, the replacement of the bitumen covering on the roof of the reactor with a noncombustible material. Many questions, however, had never been addressed, such as how it was possible to build the next two reactors when the process of decontamination in the area had not been completed. The discussion reportedly proceeded as follows:

QUESTION: Has an economic appraisal been carried out of the costs for the construction of the third stage in the changing conditions following the accident at the fourth unit?

ANSWER: Such an appraisal has not been carried out by the Atomic Energy Planning Institute.

QUESTION: How in such conditions will live and work the people who are creating the construction base, carrying out the supply of equipment and the storage of cargo?

ANSWER: Up to now, the Institute has not occupied itself with this question.

Subsequently, several participants gave their opinions about the expansion plans, including A.M. Grodzinsky of the Ukrainian Academy of Sciences, N.M. Amosov, the Chief Engineer of Chernobyl nuclear power plant, and numerous scientists of various specialities. Their conclusions were almost unanimous: that the existing plan to expand Chernobyl was now outdated because it had not taken into account the new situation that had arisen after the accident. In particular, the whole area remained dangerous, especially around the "ruddy forest," where there remained long-lasting radioactive isotopes.

One of the speakers, Aleksandr Alymov of the Academy of Sciences of the Ukrainian SSR, reportedly stated that it was no longer viable economically to build the new units at the station. He foresaw difficulties with the amount of cooling water that could be made available from the Prypyat River. This was an argument that would not have surprised the assembled scientists. But the discussions soon took on a completely different and surprising turn. Thus speakers considered that the units should not be built because their construction ran counter to existing legislation:

....according to the statutes of radiological safety, any production activity, with the exception of emergency rescue work, is prohibited in zones with an enhanced radioactive background.[40]

Another speaker at the conference, Vitalii Chumak, Director of the Center for Radiology and Ecology at the Institute of Nuclear Research in the Ukrainian Academy of Sciences, also pointed out that during the building work, contamination could be carried from one zone to another by the workers. Instead, he advocated a "forbidden area" around the nuclear plant, so that nature could recover from the devastation it had suffered after the accident.

There were also "psychological problems" to be considered. The building collective that had worked originally on the third stage had practically disintegrated and there were now said to be difficulties manning the two units in operation. They were working by shift method, but, the discussants noted, such work methods could have a negative impact on the quality of building and the safe exploitation of the station. The "periodic character of the work generally lowers personal responsibility" and people also became tired more quickly when working in stressful conditions. The matter was put forcefully and bluntly by Amosov:

It surprises me that the question is even being raised—whether or not to build the third stage. Considering how tens of thousands of people endured a severe psychological trauma and are from the top downward struggling for their future, is it morally justifiable to subject them to new doubts and fears? And these fears are not unfounded....Is it necessary to hurl a new slogan at the heroes: "Go and build the third stage regardless of the danger"?[41]

Besides, others added, such heroes had to be paid for their bravery, and the procurement of the safety of the area would also require heavy expenditure, such as taking a massive amount of contaminated soil out of the building area. Grodzinsky maintained that those who wanted to build units 5 and 6 had not learned from the experience of Chernobyl. The prevailing view in science, he observed, is that "experience gives rise to confidence," but in this case, in his view, experience led only to grave doubts. When the meeting came to its conclusion, only 2 of the 60 delegates voted in favor of building the new Chernobyl units. This was not an official decision, since Kiev's scientists did not have the final say in the matter, but it was evidently partly responsible for the resolution to postpone work on the units made in Moscow shortly afterward.

Thus ended one of the most remarkable meetings to take place in Ukraine after Chernobyl. Although the meeting was described as a "public discussion," it was not public in the Western sense of the word. In fact, it was a convocation of Ukraine's scientific elite, which, as an example of the progress of glasnost, was being permitted a role in the republic's energy development plans. And the result was an almost unanimous vote to abandon Chernobyl-5 and Chernobyl-6. The Ukrainian scientists had made their point firmly and not a little angrily, frustrated at the sacrifices that were being made in the name of progress.

Clearly the above arguments were enough to halt work on Chernobyl-5 and Chernobyl-6. But their implications went much further than cessation of work on the future units. In effect, what the speakers had revealed was that the restarting of the station itself in October 1986 was illegal, because it had occurred in a zone of increased radiation background. In addition, the above arguments were more than enough to condemn to oblivion any thoughts of returning Chernobyl-3 to service because radiation levels there were even higher than at the proposed fifth and sixth units. Every argument put forward at the Kiev meeting in March 1987 applied equally to Chernobyl-3 in December. The only possible means of response to them would have been to say that the Soviet statutes, on which nuclear power is operated, were at fault for overestimating the danger involved and that the Ukrainian scientists were exaggerating these same dangers. After the Chernobyl accident, this would hardly have been a wise course to take.

The fact remains, however, that although technically illegal, three Chernobyl units were now operational, including one in a zone of high radiation. In other words, the Soviet nuclear authorities were prepared to flout their own statutes because—if Komarov's accounts are to be believed—the Soviet Union desperately needed the power that would be produced by the Chernobyl station. Thus whereas the average load of a Soviet nuclear station is around 70%, Chernobyl 1 and 2 were run at 97% of peak capacity for three months, until unit 2 had to be shut down again.[42] Further, the radiation levels at the plant, which had been downplayed by virtually every scientist on the spot, both from the USSR and the West, were sufficient to ensure the removal from the scene of both Pozdyshev and his Chief Engineer, Shteynberg, by the spring of 1987. Both had received their "rems" and could not work within the vicinity of reactors again in their lifetimes.[43] Whether the RBMK-1000 had been made into a safe reactor is a moot point (see Chapter 7); that the area remained dangerous for personnel spending any length of time there was not in question.

There is an apparent contradiction in the above information. If the Soviets were not concerned unduly with their own laws on radiation safety, then why were the plans to build units 5 and 6 abandoned? Was it a result entirely of the shortage of water? According to the report in *Moscow News*, such "conveniences" would not normally exhaust the pros and cons concerning reactor sites.[44] The answer lay partly in the housing and accommodation situation in the zone. The evacuees remaining in Kiev Oblast had been dealt with satisfactorily on the whole, but those who continued to work by shift method in the 30-kilometer zone faced a situation that eventually became nightmarish, despite promising beginnings.

First, however, we will deal with the question of the future of the two largest cities in the area of the damaged reactor, Chernobyl and Prypyat, which between them accounted for over 55,000 residents at the time of the accident. Subsequently we will examine the partial repopulation of the 30-kilometer zone. Finally, we will analyze the plans and development of the new sites for plant operatives and building workers at Zelenyi Mys and Slavutych.

Prypyat and Chernobyl

The two major cities of the Chernobyl Raion experienced different fates both at the time of the accident and in its aftermath. Whereas Prypyat was evacuated on the afternoon of April 27, the threat to Chernobyl was not immediately apparent, and the removal of its citizens took place between May 2 and May 6, 1986. During the cleanup operations, the fact that Chernobyl was a port tempted the Soviet authorities to keep the city functioning to some extent. The Government Commission was housed in Chernobyl. It was not "clean" in terms of radiation fallout, but it served as a center of communications. Although Prypyat, as the reactor city and the largest of the two centers, has been the focus of the major attention in the West, the Soviet authorities, recognizing that the degree of contamination was much higher in Prypyat, soon began to think about a return to something resembling normal life in the raion's capital.

Following intense decontamination work, A.A. Shekhovtsov of the USSR State Committee for Hydrometeorology and Environmental Control declared in mid-December 1986 that the situation in the special zone had stabilized. He stated, to the surprise of many Western observers of the Soviet scene, that there were already 3,500 people in Chernobyl city, working on the cleanup campaign by the shift method.[45] By the following spring, there were more reports about the vast improvement of conditions in the city. No one was using respirators any more, commented Dmytro Vasylchenko, the Chief Engineer of Kombinat's Radiation Monitoring Administration. Transport and public services were said to be functioning, there was a theater in operation and a foodstore. The latter, which would have raised the most questions under the circumstances, was said to be unnecessary because "everyone in the zone gets free meals and they are delicious."[46]

At this time, Ye.I. Ihnatenko, the Director of the Kombinat association, said that the radiation background in the city had fallen "drastically." The association was looking for various new premises in the city for its workers, but was unwilling to move into the vacant private homes because of the likely return to the city of their owners. Asked about the schedule for such an event, he responded that the radiation levels would be quite permissible for normal living conditions "by Autumn." He was reminded by the interviewers of a warning that had been given by Leonid Ilyin, Vice-President of the Soviet Academy of Medical Sciences, that there should be no haste in repopulation of the city and that everything would need checking and rechecking. Ihnatenko's response was that people were already returning to various settlements in the zone of their own accord, without permission "and don't want to go away again."[47] Thus, in his view, the people would decide their own fate.

According to another source, workers were coming to Chernobyl from all parts of the Soviet Union. They were working on the reconstruction of the city's heating system and canalization, and on the building of huge hothouses. The shiftworkers were permitted to stay in Chernobyl for one month, but "in a nutshell, conditions are normal." The final part of the contaminated debris was being collected, and today, the city posed "practically no danger" to the health of the personnel living there, according to Kombinat's spokesperson, O.P. Kovalenko.[48] However, Kovalenko's remarks were directed toward a foreign audience rather than Soviet citizens. They were accurate only insofar as there could be no danger provided that cleanup workers were removed once they had accumulated the maximum permissible dose of radiation, which may have been the 25 rems discussed earlier, or the 75 rems used in times of emergency. As other sources admit, Chernobyl was by no means safe a year after the disaster.

In fact, Chernobyl had been divided into two zones: a living zone and a dead zone. The former, which occupied the smaller part of the city, had indeed been decontaminated, and people could officially move around without respirators. There, the shiftworkers consisted mainly of cleanup staff. Of the former residents, only 100 of the more than 10,000 had been permitted to return to the city by mid-May 1987, or 1% of the total. The latter area, which consisted of small private buildings and garden plots, had thus far been ignored even by dosimetrists. "No one lives there." It was stated that the main concern of the cleanup workers was not with the city's dead zone, which was not yet a priority, but rather with the roof of Chernobyl-3, 22 kilometers to the north.[49]

By the summertime, foreign journalists were permitted to visit Chernobyl, and the trial of the former Director and Chief Engineer of the Chernobyl station was also held in Chernobyl city in July 1987. The symbolic stork made its appearance, foretelling the return of residents. By June, there were comments about the buzz of the Chernobyl marketplace, with its vendors and buyers. More than 7,000 shiftworkers were evidently residing in the city, most of whom were concerned with the decontamination of Chernobyl-3. But what of the city itself. When would it be finally cleaned up for the return of its residents? A Deputy Director of Kombinat, Oleksandr Hasenko, stated that "Chernobyl desires to be and must be brought back to life." The association was currently delivering gas and hot water, and repairing the underground communications systems. But its work was being hampered by its lack of knowledge about the latter. The staff were looking for local specialists who would be more familiar with past work of this nature in the city.[50]

But was the city really *safe*? How were work conditions there? Was it not the case that the authorities were exaggerating, somewhat grossly, the return of Chernobyl to normality. A question combining the above was put to Hasenko. He replied:

I understand the context of your question. To talk about returning here for permanent residence is premature. But one can work in constant shifts. There are no health worries...[51]

In short, therefore, it seems that despite the implications of some of their comments, it had become evident to the authorities that it was still too early to think about the repopulation of Chernobyl. The apparent change of attitude on this subject seems to have occurred during April 1987, the period when the future of the fifth and sixth Chernobyl units was being reviewed, and "more conservative" forces came to the fore which demanded caution in trying to revive the zone too quickly. For the present, these forces had achieved a small, albeit temporary, victory.

Prypyat was a sore point. There is no question that the authorities were unhappy about abandoning this city of the 1970s, built specifically for the Chernobyl plant's personnel. Initially, the forecasts for the future were quite bright. Early in November, one Soviet source specified that there was "hope for a possible return to normal life for Prypyat." Another source maintained that eventually the city would be used to accommodate those workers constructing Chernobyl-5 and Chernobyl-6, who would be living there while working on shifts.[52] The comments came amid speculation that Prypyat would become a "dead city." By December, several apartment blocks in the city were said to be prepared to take in shiftworkers,[53] but there is no evidence that they actually did so at this time.

Officially, the summer of 1987 was set aside as the time when a firm decision would be made about the city of Prypyat. Meanwhile, the Ukrainian press covered the city as best it could, commenting, for example, on the wonderful strawberries that were being grown in the hothouse there, the radiological laboratory known as Kompleks. There and at other places, "hundreds of people" were reportedly working. The city had opened a swimming pool and a sauna.[54] Ihnatenko stated in May 1987 that parts of Prypyat were already clean enough for habitation, but that the territory surrounding the city was still in the contaminated zone. When decontamination work had been completed, the question of settlement there would be raised—but only about the settlement of "command personnel" dealing with cleanup work, not of families.[55]

Most Soviet sources were reluctant to talk about a ghost city. Some writers recalled what Prypyat had been like in the past:

I recall how brilliantly green this part of Prypyat once was; now it's covered with sandy gardens and asphalt (which is poured over the city, the roads and the territory to prevent dust from rising). The only sign of life in Prypyat are the hothouses, where fruits and vegetables are being raised in the hope that radiation-free produce can be grown.[56]

Others would dwell on the meager semblances of life that remained in the deserted city. Thus whereas the above writer declared the hothouses to be the only sign of life, another would use them as evidence that there was still a possibility of a return to normal life in the city:

Some of the foreign colleagues [journalists visiting the city] want to shape a sensational picture of a "dead city." But uninhabited homes do not present the whole truth about Prypyat. People are working there, an intense battle is underway toward a return to normal life. Here, for example, are large squares covered with hothouses.[57]

Mikhail Umanets is also evidently among those who anticipate that Prypyat can be restored eventually. He stated in June 1987 that one-third of its area had been decontaminated and that the "work continues and will be completed."[58] But the so-called "sensational" pictures being drawn by foreign correspondents—those who had visited the city personally—were quite accurate. Even Oleksandr Kovalenko, the enthusiastic Kombinat spokesperson, could not foresee any sort of life returning to the city for 10-15 years. Any thought of residence there had been "dropped" for the time being.[59] What Western reporters saw was a "ghost town," in which "garments still hang from windows." The city was "a silent monument to the Chernobyl catastrophe."[60] No Soviets would state outright that the city had been abandoned, therefore the foreign correspondents simply drew their own conclusions.

In mid-July 1987, the first detailed portrayal of the desolated city in a Soviet source appeared in *Moscow News* (although the newspaper enjoys a higher circulation in the West than in the Soviet Union). It provided a contrast to other Soviet works which were at pains to highlight every facet of life in the city, however insignificant:

Today Pripyat is a ghost town. There is sand at the side of asphalted roads and on paths; the upper soil has been scraped off, carried away and buried. All outer walls and balconies of buildings and their roofs have been washed with powerful jets of water. Street lamps are on at night, but you can't see any illuminated windows. There is a volleyball net in the pine grove outside a fire station. Last time they played there was the evening of April 25, 1986....At night Pripyat is deserted save for guards at the gate and emergency servicemen on duty.[61]

Aside from reasons of convenience, why would there have been hopes of restoring Prypyat to life? Would it have provided convincing proof of the success of the cleanup operation? As noted earlier, the failure to evacuate the city for 40 hours was a source of continuing embarrassment to the Soviet authorities. Leonid Ilyin has claimed that its citizens were never in any danger from radioactive fallout even at the height of the crisis. Yet, 18 months later, the situation had been put into perspective. Prypyat had indeed died as a result of the disaster. It was still unfit for longterm human habitation. Yet only 3 kilometers to the south, two 1,000 megawatt units were generating electricity. This fact was the real dilemma of Prypyat and why there was almost a desperation about resurrecting the reactor city. For how could there be justification for restarting the plant only 400 kilometers from the disaster site, when the city, further away, could not be repopulated?

Repopulating the Special Zone

By September 1986, of the reported 179 villages evacuated from the northern part of Kiev Oblast and the southern part of Gomel Oblast, only two had been repopulated: Cheremoshna and Nivetske. Both were located in Ukraine, outside the Chernobyl Raion, in Poliske, to the west of the nuclear plant. For the time being, prospects for the repopulation of the other villages did not look promising. Toward Christmas of 1986, however, Hryhorii Revenko, First Party Secretary of Kiev Oblast, announced that some centers would be repopulated during the winter or by the early spring of 1987. Fourteen villages, he stated, were being made ready to accommodate their former populations, and a further 8 villages would follow in due course.[62]

In order to prepare for resettlement, a formidable number of tasks were undertaken. Most important, the very nature of the farms had to be changed. The collective farms, stated Revenko, would be replaced by state farms and would be occupied with stock raising and feed production, as opposed to the cultivation of cereal grains. As for the homes themselves, it would be necessary to change the roofs, ensure that the water was fit to drink, and organize school, hospital and commercial services for the population. The Kiev Secretary did not think that all the former residents would return. He thought that young people, in particular, would probably prefer the new housing built for them in the southern part of the oblast, whereas the elderly would elect to return to their old homes at the first opportunity. There was no indication from his remarks that fear of radiation would be a factor in inducing people to stay away from their native villages.

By the end of the year, 12 villages in the Bragin Raion of Gomel Oblast, in the Belorussian SSR, had been repopulated. Altogether, about 1,500 people returned to the zone. A. Prokopov, the Chairman of the Executive Committee of the Bragin Raion Soviet, commented that work toward the resettlement had been carried out in several stages. First, a special committee had been established by the Gomel Oblast Executive Committee, which included people prominent in various services, such as health protection, sanitary and epidemiological services, trade, and commercial stores. Committee members then visited all those villages that had been prepared for the return of residents, and studied those aspects of village life with which they were familiar. Health was clearly the major concern. The committee then concluded that 12 villages of the raion, including Hden, Karlovka, Lyudinovo, and Paseka could be repopulated.[63]

Nevertheless, conditions in these villages were hardly ideal. Shift-work was introduced initially, but there was a shortage of medical personnel and commercial workers. Because of the severe winter, there were difficulties in obtaining fuel for the villagers, and thus coal and peat-briquettes had to be delivered directly to the homes. The Bragin Raion government was obliged to keep a careful watch over life in the 12 villages and each one was visited daily by a member of that government during the first weeks after repopulation. This early development in the repopulation process was surprising. Possibly the idea was to begin developing the new structure of farms well before spring arrived. Nonetheless, moving people back into the zone in the middle of one of the worst winters in Soviet history was a questionable maneuver. On the Ukrainian side of the border, events moved more slowly.

One reason for the relative tardiness was the concern over the spring flooding. I.S. Plyushch, Chairman of the Kiev Oblast Executive

Committee, stated in April 1987 that several Ukrainian villages in the Chernobyl zone were now ready and safe for the return of their former residents. However, since these villages were adjacent to the flooded areas, it was considered appropriate to wait until the flood had subsided and an analysis of the safety of the agricultural land could be made.[64] However, progress was very slow, even though reports about the land being used again for agriculture (see Chapter 2) were now appearing frequently in the pages of the press.

By late April 1987, two more Ukrainian villages had been repopulated, and 22 were said to be almost ready for the return of inhabitants.[65] Ihnatenko, who provided this information at a news conference organized by the Soviet Foreign Ministry, did not indicate the names of the two villages in question. One, however, was almost certainly Kupuvate, which is located about 18 kilometers north of the shift settlement of Zelenyi Mys, on the Kiev Reservoir. By mid-May, groups of farmers had returned to several other villages in the same area, notably Hubyn to the south and, across the water, Teremtsi and Paryshiv.[66] The latter is separated from the city of Chernobyl by only a stretch of water, and was thus located well inside the special zone.

The number of villages preparing for the return of their former populations eventually rose to more than 50, although it was not always clear whether the citizens had returned of their own accord before the official restoration. Konstantin T. Fursov, a Deputy Chairman of the Kiev Oblast Executive Committee, informed Western media representatives in mid-June 1987 that decontamination work had stopped completely in 27 cities and villages, for which there was no prospect of repopulation in the foreseeable future. Among the 27 was included Prypyat,[67] but the names of the other villages were evidently not provided. If they had been, it would have been possible to discern accurately the areas of most serious radiation fallout.

It is possible to draw a fairly crude division between those villages still dangerous and those that had been cleansed over the previous 15 months. Among the villages that remained in the danger zone were undoubtedly the following: Prypyat, Kopachi, Nahirtsi, Leliv, Novoshepelychi, Stari Shepelychi, Yasiv, Yaniv, Semykhody, Pidlisne, Chystohalivka, Krasne, Kosharivka and, probably, Benivka. Those repopulated (in one case never depopulated) seem to have been initially in Poliske Raion, and in the southern part of Chernobyl Raion, and later, on the eastern side of the Prypyat River. It is probable that Chernobyl itself has been included among those cities that will eventually be repopulated. The third largest city in the raion, Hornostaipil (its population is less than 2,000) is now located in the safer region and almost certainly functional.

Ironically, whereas the authorities in Moscow were dissatisfied with the demarcation of a 10-kilometer zone around the reactor by local authorities in late April 1986, and increased it to 30 kilometers, the reverse has occurred with the process of repopulation. Those villages between 10 and 30 kilometers now have a reasonably good chance of continuing their existence, whereas for those within the 10-kilometer circle, there is no such certainty. The division, however, is not the most significant point to be made about the reevacuation. The most important comment is that it occurred at all within the brief timespan of a year or 15 months. As noted earlier, there were no viable economic reasons for repopulating the agricultural workers, who could have been put to work elsewhere in the republic. It made no difference to the Ukrainian economy whether or not the villages in the zone were producing crops again, although it may have satisfied some of the villagers themselves. Prypyat would have been a different matter, but steps were already being taken in the summer of 1986 to deal with the accommodation problems of plant operatives. They, at least, would not live permanently in the special zone once more.

Zelenyi Mys

In July 1986, it was announced in the Soviet press that a new town was to be built for shiftworkers at the Chernobyl plant, close to the seventeenth-century village of Strakholissya on the western bank of the Kiev Reservoir at the very southern part of Chernobyl Raion. The location was on the border of the special zone. The town was named, after its location, Zelenyi Mys (Green Cape or Peninsula), and it was to house 10,000 workers, although the families of these workers would be accommodated elsewhere.[68] For a brief time, interest in Zelenyi Mys appears to have been high, and given its proposed size, it seems that it may have been intended originally as a permanent replacement for the deserted Prypyat. But it was plagued with problems from the outset.

During August, as building work began near Strakholissya, the work was said to be delayed by poor planning and carelessness. Progress was declared to be "inadequate." The builders were enthusiastic but were not being supported by sound organization. The workers were being housed in tents for their two-week sojourn in the area and this also could only suffice as a short-term solution for the pre-Winter period.[69]

Various high-level visits to the building site began, beginning on August 7, with the arrival of Borys Kachura, a prominent Secretary in the Politburo of the Central Committee of the Communist Party of Ukraine. On the following day, TASS announced the visit of Soviet Premier Ryzhkov and KGB Chief Chebrikov to Zelenyi Mys. Toward the end of the month, Shcherbytsky and Lyashko also went there.[70] The net result of all these visits was the cajoling of and pressure upon the builders to have some units ready for occupation by October 1, when the Chernobyl plant was to come back into service.

In order to provide workers with improved accommodation during the building of Zelenyi Mys, the Soviet authorities decided to use river housing. According to the Chief of the River Fleet of the Ukrainian SSR, N. Slavov, heavy motor vessels capable of carrying large numbers of passengers were sent to the Kiev Reservoir at the mouth of the Teteriv River from the lower sections of the Dnieper River, and from the Volga and Kama Rivers. Because the river in the area was too shallow for the vessels, 5 suction tube dredges were brought in and an area of about 7 hectares was excavated, filled in hydraulically and then a new embankment was made. A road had to be constructed that was large enough to take both passengers and freight from the embankment to Zelenyi Mys.[71]

By October 1986, about 3,000 Chernobyl operatives and builders were living on the ships while the latter worked at the Zelenyi Mys site. No doubt the vessels were warmer and more comfortable than the tents or pioneer camps that had provided earlier accommodation, but they were far from satisfactory. The main problem was that the 15 ships that were used to house the workers were not built for so many people using them all the time. There were 2 workers to every single bunk. Thus when one worker came off his shift, he slept in the bed that had just been vacated by his replacement on the following shift.[72] Life on the so-called White Steamer—all the ships together had formed a kind of floating village—was uncomfortable. According to *Trud*:

> The rhythm of the shiftworkers' life is uneasy and difficult. Whomever we spoke with, they all understand that this is a temporary, enforced matter and that sooner or later things will sort themselves out.[73]

But for reasons that may have had more to do with the haste of the project than incompetence in various areas, the development of Zelenyi Mys did not take place as planned. The ultimate plans were to build

20 "microregions" for the 10,000 workers, the first 1,500 of whom were to move in by September. One, two and three-roomed apartments were built, with a refrigerator, "modern furniture," and—what was regarded as something of an extravagance—a color television inside each one. The Housing Director at the site, T.A. Potrimai, emphasized that Zelenyi Mys was intended as a base town, and not as a hotel. Thus it was to have a village center, a pedestrian walkway, a House of Culture, a medical service, a large park—to be created from the adjacent forest—a movie theater and rest facilities, in short, "all the functional assets of a small city."[74]

The building work, however, dragged behind schedule. By mid-October, a situation developed during which the eventual abandonment of Zelenyi Mys was foreseen, less than three months after the new settlement had been announced officially in the Soviet press. The reasons were to do both with work on the town itself and the dissatisfaction of those who were living there or about to be moved there.

Once the Chernobyl station had been restarted, there was an immediate need to accommodate more personnel. The sarcophagus had to be completed, and preparations for the startup of units 2 and 3 were under way. A massive area had to be decontaminated quickly before this process could take place. Thus workers began pouring into the new settlement in October. They tried to stay on the White Steamer, where the food was said to be good and conditions were "decent," but the liners were already packed to capacity. Zelenyi Mys could only accommodate 850 people by October 1986. Some 400 apartments had been completed, but had no water and therefore no heat either. When the water was switched on, the plumbing had failed. Damp appeared through the wallpaper and floorboards, even on outside walls. The move into Zelenyi Mys had to be postponed; "the happiness about the move was badly clouded."[75]

In addition, the town did not have a decontamination center, the most important item in a settlement for operatives working in a zone of increased radiation for 12 hours daily. This, as *Trud* pointed out, should have been built first of all. Therefore nuclear power specialists were warning that plant personnel could not be moved into the town until the center had been built. The newspaper's advice was all but ignored, however, as the 850 people were living there already without such a facility. Even in December, there were complaints that sometimes the heat in the houses was switched off, at others there was no water available. As for the subcontractors, hired by the trust "Pivdenatomenerhobud" [Southern Atomic Energy Construction], retributions were eventually given out for their defective workmanship.

In early October 1986, the Ukrainian newspaper *Robitnycha hazeta* had sharply criticized the Volgograd Wood Processing Plant called Kuibyshev for sending wooden houses to Zelenyi Mys that had defective sanitary engineering. The enterprise's party committee had subsequently called a meeting and handed out strict reprimands to the enterprise's Director, V. Koldayev, and its Chief Engineer, V. Petrov. The Secretary of the Party Committee was rebuked for failing to ensure the timely delivery of high-quality goods for Chernobyl workers. The Volgograd Plant's Chief Technologist, Chief Designer, Deputy Director and Material-Technical Supply Director were also "brought to strict party responsibility" for the shoddy nature of the homes provided. Within the plant itself, the Deputy Chairman of the house-building shop, S. Pogulyai, and the Senior Foreman, L. Goncharov, were dismissed. Other senior personnel had money deducted from their wages.[76]

The Zhytomyr furniture makers had also send furniture with defects to the new settlement town and various personnel from the supply organization had visited the site in order to correct the problems. It was "pointed out to them firmly" that furniture had been sent to Zelenyi Mys but that the quality control over its manufacture had been at a low level.[77] Long before these retributions took place, the future of Zelenyi Mys was called into question. The first hint that something was seriously wrong came from Erik Pozdyshev, in his capacity of Director of the Chernobyl station:

There is one problem that keeps cropping up. People are asking—why is our accommodation allotted on a temporary basis? After all, we are permanent workers, why are our apartments and residence permits temporary? What is the reason? These questions have been raised at all meetings and in party groups, and quite honestly, I have to say that I cannot think of a sensible answer to this question....The point is that housing for personnel is issued in Kiev and Chernigov [Chernihiv] and the temporary residence permits are also issued there. It is true that we will build a new city, a decision to this effect has been taken, but it is a matter for the future. So why this distrust now? Some people claim that if we allocate permanent apartments, people might leave the atomic energy station....No, I do not believe in this approach. Those who have remained at the station have weathered a difficult test—who has the right not to trust them?[78]

Pozdyshev's remarks indicate not only that workers wanted something more permanent than a shift settlement located on the edge of the special zone, but also that they felt they were entitled to permanent accommodation for themselves and their families. Once the station itself had restarted, it was evident that their future employment in the area was ensured, but they still had temporary permits. Their predicament was compounded by the low quality of the housing provided at Zelenyi Mys. Therefore they made known their complaints and the consequence was that the authorities resolved to provide something more permanent for them. The decision may have also been influenced by the fact that the future of Prypyat appeared increasingly uncertain.

As a result of the above developments, on the day that Zelenyi Mys had its official "housewarming," it was also announced that it would not be a permanent settlement for Chernobyl plant operatives. For the time being, it was reported, it would continue to be the main location for the duty brigades at the station. But when Slavutych was completed, then Zelenyi Mys would become a recreational center (this announcement came only two weeks after the comment noted above that the settlement was a base and not a hotel). Its ultimate size was reduced from 10,000 to 5,000, but by mid-December 1986, only about 3,000 residents were actually based there.[79] There was a complaint in the summer of 1987 that there was far more dormitory space at the settlement than was warranted by the number of people who would be living there, and that these same dormitories should have been built at Slavutych.[80]

A Western account about the shift settlement at Zelenyi Mys described it as an "ugly, prefabricated settlement" and "a curious cross between a wartime camp and a holiday retreat."[81] Photographs and Soviet reports bear out this description. The town "looked" temporary. It was obviously too close to the damaged reactor to house comfortably the thousands of plant operatives who would be working at Chernobyl. Moreover, the disruption of family life that occurred after Chernobyl had probably continued too long for the workers involved. New settlements, after all, had been constructed for those evacuees who could not return home. They alone had temporary permits. There was also a distinct difference in the authorities' attitude toward builders and that toward plant operatives. The feeling—and this emerged with full force at Slavutych—was that the operatives should be provided with the best accommodation available, whereas the building workers' needs were less important.

The outcome was that Zelenyi Mys, which at first was a source of great anticipation among the Ukrainian party hierarchy, was soon relegated to a secondary status, condemned partly by its failure to meet the needs of the workers. Unfortunately, the problems there were mild by comparison with those that developed at Slavutych.

The Slavutych Saga

The first official announcement of the construction of the city of Slavutych came in late October in the Soviet press. Pozdyshev, during his interview in *Pravda* on October 10, 1986, had mentioned that a new city was to be built, but he had considered it a "matter for the future." Yet it was scarcely two weeks after his comments that the location and name of the new city was revealed. Its location was on the Dnieper River, some 45 kilometers to the northeast of Prypyat, on the territory of Chernihiv Oblast. At the site was a "forgotten railroad station" called Nerefa, while the nearest village on the map of Ukraine was Nedanchichi, almost on the westernmost point of Chernihiv Oblast in an area of forest, remote and virtually unpopulated. The location had two advantages, one natural and one man-made. It was on the Dnieper River, which enabled the transportation of supplies and, as will be shown, the use of boats for the temporary accommodation of the building workers; and there was already railroad access to the area, so that a linkup could be created from Slavutych to the Chernobyl nuclear plant.

The plans for Slavutych were first devised in Moscow by the Central Scientific Research Institute of City Construction, based on research undertaken by the all-Union and Ukrainian Institutes of Engineering-Technical Research.[82] Control of the design was then handed down to the Kiev Zonal Scientific Research Institute of Standard and Experimental Planning, which was given the job of putting together the final design in only about 6 weeks as compared to the average of 18 months usually required to design a new city.[83] The Kiev Institute was assisted by 34 planning organizations from 8 Union republics: Armenia, Azerbaidzhan, Georgia, Latvia, Lithuania, Estonia, the Russian Republic and Ukraine. The official client for Slavutych, the future city for Chernobyl plant operatives, was the Ministry of Atomic Power Engineering of the USSR, while the official contractor was the Ministry of Power and Electrification of the USSR.[84]

Slavutych was planned as a "21st century city" with every modern convenience. The original plan anticipated that it would be occupied by the Fall of 1987, and that its population at the end of that year would be 10,000, rising to 20,000 by the end of 1988, and ultimately to 30,000 upon completion.[85] It was to be a city built of white stone, it was announced, comfortable, beautiful "and above all, green." The Chairman of the First Architectural Planning Workshop of the Kiev Institute, F.I. Borovik, described it as follows:

> It will not only be surrounded by coniferous forest with impregnated, leaf-bearing trees, but it will also preserve the natural environment to the utmost in each of the four residential complexes. The planners have tried to revive the partly lost concept of the pedestrian street: these are being planned in each separate section. These will be adjoined by schools, stores and service areas. At the center will be a market-square with a statue of V.I. Lenin, the building of the City Party Committee, a theater and a House of Culture.[86]

The city was to be based on multistory buildings, including a 9-floor guesthouse. Seven preschool establishments were planned for 280 places each and 3 kindergartens. Close to the center of the city was to be a museum dedicated to the heroes of Chernobyl. Each Union republic had put forward its own plan; they were divided up into five distinct construction areas and were encouraged to develop their own distinctive style in each section, within the general limits of the architectural plan. One section was under the charge of the Ministry of Power of the USSR, the second was run by the Ukrainian SSR, the third by the Russian SFSR, the fourth by the Baltic republics, and the fifth by the Transcaucasian republics. In 1987, the goal was to complete several residential buildings, 3 kindergartens, a school, a health complex, a cafe and a restaurant, stores, a public bathhouse and a hotel.[87]

In the Ukrainian sector, the chief organizations involved in building the city were those of the Ministry of Construction of the USSR (the largest of which was the trust Slavutych Atomic Energy Construction), the State Agro-Industrial Committee, the Ministry of the Coal Industry of the Ukrainian SSR (which was abolished and replaced by the all-Union ministry in August 1987), and the Kiev Urban Construction Division.[88] Builders involved in these organizations were to remain in Slavutych for two years, along with colleagues from the other seven Union republics.

Slavutych, clearly, was a different proposition to Zelenyi Mys. It was conceived as a major project, the largest building project to emerge as the result of the Chernobyl disaster. The future city was to cater to those based at the expanding Chernobyl plant, along with their families. Consequently, those building the city had to adhere to a timetable much more rigorously than builders of other cities in the USSR. We have seen that on October 1, 1986, the Chernobyl plant came back on-stream even though the covering over Chernobyl-4 had not been completed. This demonstrated that where nuclear power engineering was concerned, the harnessing of new electricity sources took priority. In other words, the new Chernobyl reactors would be operational whether or not the new city for operatives was ready. However, the operatives were already complaining about the lack of facilities and the hardships in Zelenyi Mys. They were hardly likely to be satisfied to hear that there had been delays in building the new city, which would allow them finally to be reunited permanently with their families and to lead a normal life again.

Unfortunately, matters did not go smoothly with Slavutych, despite the widespread publicity about the new city in the Ukrainian press, and the optimism about the attractive new city that would provide a worthy replacement for Prypyat, now abandoned in practice, if not in theory. The major problems lay with the building workers and particularly their accommodation in the area. In the Fall of 1987, everything appeared to be proceeding according to plan, but then by the spring of 1987 had turned into a crisis situation.

At first, the location of the new city only 10 kilometers from the Dnieper River enabled the authorities to establish a floating settlement, called Yakir, for the building workers. Eight dredges had been brought up the river from Kiev and Mozyr to deepen the base of the river. Crews worked around the clock pumping sand slurry from the bottom of the river and in about a month, the channel had been deepened over a 40-kilometer section from the village of Teremtsi in Kiev Oblast to Nedanchichi. This not only enabled the use of ships for accommodation in what had been shallow waters in Chernihiv Oblast, but also allowed for the transport of heavy cargo along the river route. Six diesel and electric-powered ships were tied up to the river banks to house the workers and specialists. Each could house 200 people. By the winter, a further four ships had been added, so that about 2,000 workers and specialists could use them for meals and overnight accommodation.[89]

On the ships, electricity and telephone communications were provided and drinking water was obtained from "200 deep wells." Nevertheless, the ships were not intended for longterm settlement even on a shift basis, and before the end of the year, some of the builders were moved to a new shift settlement (about the same distance from the

Slavutych site) called Lisne [Russian: Lesnoye] where 1 and 2-story residential buildings had hastily been made ready for the workers, and stores were opened with consumer and industrial goods. "Brandenberg-type" dormitories had been provided for the workers from East Germany, Finland and Czechoslovakia. There were no streets in Lisne. Eventually there was to be a metro linkup from Nedanchichi (Lisne) to Slavutych that would collect the "transients" from each of the various republics in the morning and evening, but this was evidently not built in the 1986-87 period.[90]

On January 8, 1987, a high-level meeting was held in Kiev to discuss the situation at Slavutych. It was attended by Politburo Candidate Member and Secretary of the CC CPSU, Vladimir Dolgikh, Ukrainian First Party Secretary, Volodymyr Shcherbytsky, and Ukrainian Premier, Oleksandr Lyashko. Each of the Union republics working at Slavutych was represented at the meeting by a Deputy Chairman of the respective Council of Ministers. The speakers, Shcherbytsky and Dolgikh, emphasized that the "liquidation of the consequences of the Chernobyl accident" had entered a new stage, which was no less important and complex. The period of organizational work had to be reduced in order not to slow down the tempo of building work at Slavutych. The chief criticism appears to have been directed at the Ministry of Transport Construction of the USSR (and Minister, V.I. Brezhnev) for failing to complete the rail connections that would enable a rapid increase in the volume of cargo deliveries to the site.[91]

The meeting indicated that Slavutych was falling behind schedule. Shortly, it became apparent that there were some flaws in the design plan and also that the work crews were seriously understaffed. Thus A. Volkov, the Deputy Director-General of the Kombinat production association, commented in March 1987 that according to the current plan, there would only be enough living space in Slavutych for those operatives at the first and second reactor units. When planning the city, he added, no provision had been made for the future expansion of the nuclear power plant. Nor had anyone considered where the building workers themselves were to live. The prefabricated huts, which had only partial insulation against the winter cold, were obviously a temporary solution, one source reported:

For the time being, even the Ministry of Power's engineering and technical staff in Slavutych are under 25% of their full strength. There is a shortage of workers—they do not come here because they are not given any guarantee of permanent housing. It is clearly possible to build a city for the construction workers adjacent to Slavutych. But would that be viable? After all, you would then require extra nurseries, stores, hospitals and movie theaters....Moreover, there is neither the time nor the workforce available for such work.[92]

By the spring of 1987, while criticisms such as that above were just surfacing, the official status of Slavutych in the Ukrainian SSR was verified by the elections of local organs of government. Slavutych had remained something of a curiosity insofar as administration was concerned. Geographically, the new city was clearly located in Chernihiv Oblast. All the early reports about building work there were in fact signed from Chernihiv Oblast. By the spring, however, reporters at the site would sign their articles: Slavutych, Kiev Oblast. Yet since the Belorussian SSR divides the two Ukrainian oblasts directly west of Slavutych, its location in Kiev Oblast appeared to be an impossibility.

The answer to the puzzle lay in the transfer of the Prypyat City government and the Prypyat City Party Committee to Slavutych, including First Party Secretary, Viktor Lukyanenko, who became, *ipso facto*, First Party Secretary of the Slavutych City Party Committee. The city government elections may also have irritated the building workers who were dissatisfied with their transient status (and they were described as "transients" in Soviet accounts), because they were permitted to vote, but then had no prospects of ever living in the city. Extraordinary steps were taken to ensure that the plant operatives could cast their vote. Collection points for ballot boxes were established as far away as Chernobyl and Zelenyi Mys, while the operatives could vote during their work hours at the Chernobyl plant. Thus it was possible to cast a vote for a candidate in an unbuilt city at a distance of up to 67 kilometers away. Although the leaders of the Prypyat City Executive Committee seem to have retained their posts, many of the candidates in the Slavutych election were standing for the first time.[93]

The task of the new city administration was to deal with the development of Slavutych and the "numerous problems" that had developed in the social and other spheres. On March 6, 1987, in an effort to alleviate the labor shortage, a 325-man Komsomol crew was dispatched to Slavutych at the behest of the 25th Komsomol Congress of the Ukrainian SSR. They arrived expecting to receive preferential housing and remuneration, but owing to apparent misunderstandings between Kiev and the building trust Slavutych Atomic Energy Construction, headed since November 1986 by Viktor Svinchuk, the Komsomol crews were completely unequipped for the tasks to which they were assigned. They had also anticipated that they would be able to work as one unit at Slavutych, but in practice were dispersed among various work crews.[94]

Although retraining on a crash project such as Slavutych was "an impermissible luxury," the Komsomol team was given a brief course. But after only six weeks, 15 of its members had been dismissed for various reasons. Although the number represented only 5% of the additional workforce, the remainder had evidently given their notice to leave. The dilemma, according to one account, was a result of the lack of communication between the Slavutych Atomic Energy Construction trust and the Komsomol leaders in Kiev:

> We were told that currently in the Central Committee of the Ukrainian Komsomol, to the staff of the all-Union Komsomol "rush-construction" [i.e., crash building projects, requiring so-called *udarniki* or shock-troops], hundreds of letters arrive from young building workers who want to work on Slavutych. But will mistakes recur, and will the wrong people be sent? It can happen. As Serhii Zabaryn, Chief of Staff of the Komsomol Construction Department declared, there is no permanent linkup between the Komsomol and the trust leadership.[95]

Both the Slavutych Atomic Energy Construction trust and the Kombinat production association were reprimanded by the buro of the Prypyat (soon to be Slavutych) City Party Committee in April 1987. According to Second Secretary, Serhii Kostin, the work of both organizations was not up to standard. There was also a "negative phenomenon" of workers lingering in sobering-up stations after drinking bouts. Alcoholism was also impeding the building of the city, and there were said to be delays in the construction of stores, the cafeteria, the medical complex and other establishments.[96]

In late April, there were also serious domestic problems in Lisne. The cafeteria, for example, "literally" shook when there was an influx of people, even though it would eventually have to withstand crowds of more than 1,000 compared to the 250 it took in at that time. The food was said to be inadequate for workers involved in heavy physical labor. A better cafeteria was required, although Svinchuk was showing an "enviable calmness" about the entire matter, according to the Soviet source. Some of the workers had decided they would be better off taking care of their own breakfast and lunch. But the tiny store in Lisne provided only baked goods, canned fish and cooked kobassa. It did not even have any bread.[97]

The "Soyuzpechat" (Union press buro) kiosk usually sold out of newspapers and magazines almost immediately, having a meager supply—although this is a common criticism in the USSR, and also occurs in Moscow. There were no facilities for sending and receiving mail in Lisne, and the workers had to drive the 10 kilometers to the nearest postal service in Nedanchichi. The staff, it was reported, "literally groans" about the impossibility of reaching Kiev and other cities by telephone, because the Chernihiv Oblast authorities had not ensured the connection. It was even more difficult to place calls within Slavutych itself. Every Union republic work organization was said to be complaining about the lack of reliable telephone connections in the new city.

The women's dormitory was said to be flooded because a pipe had burst, but no one had bothered to fix it. The kitchen was said to be in an even worse state. The women, it was stated, were spending more time on their "hairdos and faces" than on tidying up the kitchen. The source advised the Komsomol staff in particular to have a talk with its young workers, to remind them that since they were building the city of the future, then the Lisne settlement also belonged to the future and should be a model facility. Ironically, the visitors to Lisne had unwittingly hit on the fundamental problem of Lisne: that it was not, unlike Slavutych, part of the future, but something temporary.

As a result of these reports, the Kiev Oblast Komsomol Committee decided to investigate for itself the state of affairs in Slavutych/Lisne, and organized a "raid" (that is, arrived unannounced) on the area in mid-June 1987. The raiding party, known as "searchlight operators," "discovered problems from the start." They first examined the concrete unit. The concrete factory had been shut down for 4 days through lack of cement, and the operators wondered how it was possible to check its quality in the first place given the lack of a laboratory and appropriate instruments for the purpose. The logbook of the Slavutych Atomic Energy Construction trust reported that in mid-May, the concrete mixer had broken down. The cement had to be sent from Vyshgorod and Chernihiv and "presents quite an expense to the country," but the trust did not keep any in stock and was thus reliant on the unpredictable railroad supplies of the material. Altogether, it was ascertained, the factory had worked only one half-day out of ten in May. A proper repair base was needed for those machines that were constantly breaking down.[98]

The operators next visited the "Berlin-1" restaurant, a 1-story dining hall, which was said to be almost completed. However, the room was half-empty. The manager, O.A. Samoilenko, explained that although scheduled to open in March 1987, the staff were still awaiting kitchen equipment—he implied that as it was all imported, it was being delayed at a customs post. To save time, it had been decided that rather than wait for its arrival, Soviet-made equipment would be used instead. But in order to install local equipment, the floor would have to be ripped out, cables changed and a new supply of electricity prepared. These matters would result in further delays of the dining hall.

The major concern of the investigators was the labor collective, which was lacking in both equipment and protective clothing. Workers were said to be "taking offense" at the absence of work gloves and footwear, which was a consequence of deficiencies in supply. The organization of the work itself was described by a builder from Ivano-Frankivsk who detailed the erection of a facility for the Slavutych liaison department:

> In the morning they told us: "We have to raise the level of the construction site by 30 centimeters. We poured on 30 centimeters. "Oh," they said after lunch, "there has been a mistake, we need only 10 centimeters." The next day geodesists arrived. "Men, pour on 1.5 meters!"[99]

Another source of anxiety, especially among the workers themselves, was wages. Paydays often came 4-5 days late and, more serious, there were said to be substantial irregularities in pay. The reason for the discrepancies concerned the status of workers. Those who belonged to the Slavutych Atomic Energy Construction trust were considered "local," even though they obviously did not come from either Slavutych or Lisne originally. On the other hand, those sent into the area from their regular job were said to be "on a mission" in that they would eventually return to their primary work place. Matters were complicated by the system of bonuses devised for those people building Slavutych.

Thus, the Ministry of Power and Electrification of the USSR decided that all building workers at Slavutych would be eligible for bonuses of up to 25% of the normal payrates for the job. However, for those workers "assigned" to Slavutych, and sent from their regular jobs, "normal" conditions were waived. These workers were entitled not only to the normal rate of pay and a 25% bonus as indicated, but also to 75% of their regular pay at their primary job:

> Imagine this scene. Two workers of equal qualifications work side-by-side, let us say on the construction of a boiler house or a housing

facility. They live in one dormitory, ride to work on one and the same bus, carry out the same complex physical work, breathe the same air....But for one day's pay, one receives almost twice as much as the other. Why? Because one is considered on a mission (he was really sent here by some construction organization, let us say from Kharkiv, Donetsk or Poltava), and the other was accepted for work on the basis of an announcement in Slavutych directly, even though he also arrived here from Kharkiv, Donetsk and Poltava, and he is also on a mission (he is registered temporarily in Slavutych, lives far from his family, and has no prospects of receiving an apartment here).[100]

Even the bonuses were sometimes withdrawn. Vasyl Martin of Zarkarpattya in western Ukraine, from an impoverished family of eleven, arrived at Slavutych having completed his military service. For his first month's pay, he received 240 rubles (the average monthly wage in the Soviet Union is around 220 rubles, slightly less in Ukraine). The leaders of the work brigade then decided that the men were entitled to a further bonus, which was duly distributed. However, when the second payday arrived, it was announced that the "wrong amount" had been handed out the previous month, and that the workers were now in debt to the trust. Consequently for his second month's work, Martin received only 14 rubles. He noted, ruefully, that the cost of food alone amounted to 3 rubles per day. The result of the shortfall was that "our brigade immediately fell apart." Some workers joined up with a neighboring unit from Armenia which was said to be better organized. Martin himself was considering leaving Slavutych altogether and looking for work at the Tyumen oilfield in western Siberia.[101]

Why was life so difficult in the Ukrainian work area, as opposed to other areas? Soviet sources concur that the Ukrainians were given the stiffest task in terms of the land on which they were building. One maintained that as the host republic, it was incumbent upon Ukrainians to take on the most difficult area. It was also evident that many of the workers for the various trusts had arrived at Slavutych having completed the building of homes for Chernobyl evacuees. For example, the Chernihiv Industrial Construction trust, headed by Yurii Vasylevych, had built homes for evacuees in Lukyanivtsi, in Kiev Oblast, before moving on to Slavutych.[102] The builders therefore had been laboring for more than a year for Chernobyl victims at the time of the Komsomol raid, and yet had no prospects of remaining in the new city once their term of work was over. They were irritated and often angry at the situation in which they found themselves. Moreover, when things at the site went wrong, the Prypyat City Party Committee was quick to blame the builders. Thus in June 1987, on the orders of the committee, Svinchuk "severely reprimanded" the building workers at Slavutych "for their irresponsible attitude toward ensuring the timely construction of social-domestic structures" and the unsatisfactory labor and holiday conditions.[103]

The workers felt that they had been deceived by the authorities. One builder, Ihor Sobkovych, commented that he had been attracted to work at Slavutych first because of the romantic nature of such a grandiose construction, and second, because he had been promised the opportunity to become a resident of the new city. Another, Fedir Turyk, a member of the Central Committee of the Ukrainian Komsomol, pointed out that although builders had endured much worse conditions in the past than at Slavutych—suffering from scurvy and living in remote areas in the Far East while building the railroad, for example—their prospects in terms of living conditions had at least been made clear to them beforehand:

> They were told: build a city and be its masters. And they built it. When we were persuaded to come here, we were told the same thing. But now they say: build it—and goodbye. And perhaps after the completion of the Slavutych construction, we will not be needed by the city or at the nuclear plant? But why not construct a few buildings for us here—of the hotel or even apartment type? You see, it is completely normal that young people yearn to put down roots, to find their place in life.[104]

Some of the young people attempted to "put down roots" by getting married in Lisne, which was perhaps not surprising given the fact that the average age of the building workers was 27. But the married couples then had no place to live, since there were very few double rooms in the dormitories. One couple were permitted a brief honeymoon in the local hotel, and then unceremoniously ejected from the premises. An entire brigade went on their behalf to the construction headquarters to request that the couple be given one of the double rooms in the dormitory. They were met with the response: "Today we give this couple a room, tomorrow everyone will get married." Yet at the time the above account appeared, seven more weddings were anticipated in Lisne, so the problem became worse over the summer of 1987. The couples, however, could not even get on a waiting list for the apartments in

Slavutych, since no such list had been drawn up and the city government had not yet been set up (even if it had been, there is no guarantee
that the builders would have been on the list).[105]

The outcome of the widespread discontent at Slavutych was that
some of the workforce, and in particular the more skilled workers,
decided that enough was enough, and were no longer prepared to accept
a situation that held so few guarantees for the future. Although there
are usually frequent turnovers of staff on any major building project,
rarely does it work out that more staff leave than start work. By June,
however, this was what was happening at Slavutych. At the largest of
the more than 100 organizations involved in the building of the new
city, the Slavutych Atomic Energy Construction trust, there were only
about 50% of the required laborers, and a shortage of almost 150
engineers by late July 1987. In June 1987, 41 new workers had joined
the trust, but 69 had left. Over the May-July period, over 200 trust
workers had left their posts and there was a sizeable net outflow of
engineers and technical workers.[106]

By August 1987, according to a Soviet source, there were questions
about building work and about the very status of Slavutych.[107] The
delays on various building projects warranted another visit from Shcherbytsky and the new Ukrainian Premier, V.I. Masol, on September 15.[108]
By the end of the year, there were indications that many of the operatives
who might have expected to be moved to Slavutych from apartments
in Kiev and from the shift settlement at Zelenyi Mys were in fact still
waiting to move, and sometimes none-too-pleased about the matter.[109]
It appears that the city had in fact been prepared for its first residents,
but that the central heating system had broken down.

Thus by mid-November, the builders had reportedly completed their
annual program, despite the general atmosphere of discontent and the
various problems outlined above. A middle school had been built, in
addition to a drugstore, a kindergarten and 1,200 residences. All
depended on the startup of the central heating system, being built by
the trusts Special Atomic Energy Assembly, Southern Heating Energy
Assembly and others. By using large blocks in assembly work, the time
period was evidently cut down by 4 times, but there followed the sort
of blunder that has impeded the construction of Slavutych from the
outset. The concretization of the holes of the filters in the building that
cleansed the water that would reach homes was not done properly. As
a result, the water was not purified of its various ferric oxides, gravel
and other admixtures.[110]

This mistake, it was reported, cost the Slavutych residents dearly.

The unconditioned water prevented the boilers from working at optimum capacity. At a temperature of more than 80 degrees, scum accumulated in the boilers and the pipelines causing premature wear-and-tear on the energy equipment. It was proving impossible to ensure clean drinking water for the future residents. The pipe water, according to one builder, was cleansed one day, but had turned red again by the following day. The water pressure was falling, and one pump was already out of action because of the excess demand for water. There were in fact only two alternatives: either to cut off the city's water supply entirely in order to obtain the necessary pressure of the water; or to provide the water to the city, which would put a strain on the boiler and eventually lead to breakdowns. (Even then, it is likely that there would have been restrictions on heat and water in the homes.) "And all this only because of the bad quality of the water."

In addition, some sections of pipe in the city's heating section had not been covered with reinforced concrete. Although winter was imminent, several buildings in the industrial base had not been provided with heat, most notably the concrete solution factories, of which 2 out of 3 had to operate in winter conditions. No storehouses had been provided for cement, inert materials and chemical goods. Yet, stated the report, if the factories should stop work, then the entire construction will be on famine rations. There would be neither concrete nor solution.

In late November 1987, the Slavutych City Party Committee held a meeting. Lukyanenko gave a speech, during which he berated the work of the Slavutych Atomic Energy Construction trust and of the Kombinat production association for their failure to keep to deadlines in completing the various buildings at Slavutych. They were not applying the old psychology of meeting deadlines "at any price," he lamented. The party committees were incapable of supervising such operations properly partly because they were in the process of restoring their depleted personnel, but also because the City Committee had not ensured that those involved in "liquidating the consequences of the accident" (which involved everything from cleanup work to building Slavutych), many of whom were non-professionals, were given a proper account of leadership work. In short, party work had been hindered by the turnaround in personnel.[111]

Although a considerable portion of Lukyanenko's speech was printed in the Ukrainian press, the most controversial sections were omitted. Some of these were revealed only in the published account of the Plenum of the Kiev Oblast Party Committee of the Communist Party of Ukraine, in early December 1987. They concerned not only "gross violations" by personnel at the Chernobyl nuclear power plant (discussed in Chapter 7), but also the "violations of technological discipline" that had occurred at the building site in Slavutych. According to Lukyanenko, over the past 10 months, there had been "36 unfortunate incidents," including 3 fatalities. The questions about work safety had not been uppermost, he stated, in the minds of economic leaders and trade union organizations, and the workers themselves were exhibiting very low levels of personal responsibility. Several officials were made "answerable to party responsibility," including plant Director, Mikhail Umanets and the General Director of the Kombinat production association, Ye.I. Ihnatenko.[112]

When Western news agencies picked up the news of 3 fatalities and perused the somewhat muddled account in *Sotsialisticheskaya industriya*, many assumed that the deaths had occurred at the Chernobyl station itself. The reprimand given to Umanets indicated as much. Clearly, something serious had occurred at the nuclear plant. But the Soviets insisted that the "gross violations" there had not resulted in fatalities. At a press conference held in Moscow in early December, the incidents were discussed by Yurii Filimontsev of the Ministry of Atomic Power Engineering of the USSR. He stated that the Western media had misinterpreted Soviet reports. Three deaths had occurred, but two had been at the building site in Slavutych, whereas one man had drowned in the Dnieper River, again at the Slavutych site.[113]

If the above report is accurate, it illustrates again the lamentable state of affairs at Slavutych. The first year of building the new city had been an unmitigated disaster: 3 dead, a polluted water supply, 325 Komsomol workers leaving their jobs, including 15 fired, buildings in a dilapidated state and incomplete, engineers leaving the site, wage discrepancies, and altogether an impression of a thoroughly dissatisfied workforce, as the Kiev Oblast Committee Plenum acknowledged. Why had such a situation occurred? Slavutych was a major project of the highest priority, and was supposed to be a model city, after all.

There appear to be two related reasons. The first is that the city was planned and constructed at an alarming pace. In this respect, it was typical of what happened elsewhere after Chernobyl, The situation of the plant operatives had to be rectified with the utmost speed, partly because of the need to cater to an expanding nuclear plant, and partly because Zelenyi Mys had not lived up to expectations. In fact, one can surmise that a major reason for the discontinuation of Chernobyl-5 and Chernobyl-6 reactors was that there was simply nowhere to house either the building workers who would be needed to construct those reactors, or the operatives who would be required to run them. There were already indications that the Soviet authorities planned to build another nuclear power plant close to Chernobyl on the Desna River, to the northeast.[114] This plan revealed that it was practical rather than safety considerations that had caused the cessation of building work on units 5 and 6 (as noted above, safety considerations can hardly have been the major concern given the restarting of unit 3 on December 4, 1987). The most obvious problem was housing. Slavutych was built rapidly, but also carelessly and negligently and therefore did not meet expectations.

The second reason for the problems at Slavutych was that in attempting to build the city, the authorities neglected to look after the needs and interests of the workforce involved. Had the builders been guaranteed apartments in the new city and permanent residence status, then it is likely that they would have taken a much greater interest in the final result of their work. Instead, regarded and treated as transients, their attitude was clearly one of disinterest in this 21st century city that they could never live in. Conditions in Lisne were deplorable, which made matters worse.

Slavutych will not be the last chapter in the history of Chernobyl, but it is one of the unhappiest ones. Was there another solution? The answer is that there was, and that it is related to the other topics discussed in this book. Had the authorities not been so anxious to restart the plant, to repopulate the zone, but rather to plan matters at a more leisurely pace (yet more thoroughly), then they might have anticipated many of the predicaments that arose. In the final analysis, it was because operatives were working for so long on a shift basis that Slavutych had to be built so quickly. Had the Chernobyl plant been closed down, even for a mere two years, then there would have been no need for operatives to make sacrifices in the first place.

7 The Nuclear Power Debate

The Chernobyl accident compelled many countries of the world to question their nuclear power programs. In Yugoslavia, for example, which participates in many of the meetings of the CMEA countries, there have been massive protests against nuclear power (the country has one nuclear plant in operation). In 1987, the Yugoslavian Socialist Youth Federation began a campaign against nuclear power plants, while a leading sociologist in the country stated: "That Yugoslavia did not learn anything after Chernobyl is evident in the fact that it has not discontinued its nuclear program."[1] Other European countries, such as Sweden and Austria, have also rejected nuclear power as a viable option. The Soviet reaction, however, has always been that it was man rather than the machine that was at fault in the nuclear accident, and that nuclear power remains the safest and most viable of Soviet energy options.

Nonetheless, the Soviet authorities have made some concessions since Chernobyl, not least being the fact that the Soviet nuclear industry has suddenly been subjected to international scrutiny by the International Atomic Energy Agency (IAEA). Before the summer of 1985, the IAEA had not set foot on Soviet territory. This situation changed after the meeting of August 1986 (see the Introduction), and thereafter, the Soviets have been at pains to point out that each post-accident step in the sphere of safety in the nuclear industry has had the approval of the IAEA. This sort of scrutiny, be it of short or long duration, must be seen as a positive sign. There is no question that Soviet nuclear plants today are safer than before the Chernobyl accident.

In the process of Soviet-IAEA cooperation, a new theme has been emphasized in Soviet academic literature, which is that this close relationship has a long history, including a special Ukraine-IAEA association through the aegis of the United Nations. Aside from the scientific benefits of such an association, there is also a political aspect to such an emphasis. Vienna was a public relations triumph for the Soviet Union. But in order to maintain the goodwill generated there, it was necessary to be seen to be continuing the Soviet-IAEA relationship. But how close is this relationship, and is it likely to be of long duration?

Moreover, while one can accept IAEA expertise in the field of nuclear energy, the organization is hardly objective when it comes to an assessment of energy alternatives. One does not have to be a anti-nuclear activist to recognize that the IAEA gives short shrift to the possibility of more reliance on coal or hydroelectricity, for example, in the production of electricity. In this respect, the views of the organization coincide with those of the people deciding Soviet energy policy, the latter being an indefinite conglomerate of party members and prominent scientists. In brief, Soviet assessments of the future, and IAEA blessings for the improvements made to their reactors, do not always add up to an objective analysis. All arguments about the industry aside, one fact should be pointed out at the outset: the Soviet Union is fanatically pro-nuclear power. The industry is centralized, run directly from Moscow, and all important decisions come from the center rather than the regions in which the nuclear plants might be located.

In the Ukrainian SSR, for example, which has about one-third of Soviet nuclear capacity (including the Chernobyl plant), there are no decision-making agencies for the industry. The Ukrainian Communist Party Politburo might be enthusiastic or reluctant about the nuclear energy program in the republic, but the outcome will be the same. Those officials within the Ukrainian Politburo, such as Borys Kachura, who have actively supported central policies, have seen notable advancement of their careers as a result. But the industry before Chernobyl was run by the Ministry of Power and Electrification of the USSR, with the weapons production side being under the control of the Ministry of Medium Machine Building of the USSR. Today, the civilian industry has fallen under the control of the newly established Ministry of Atomic Power Engineering of the USSR. All are based in Moscow. There is no representation at the regional level in Kiev.

Chernobyl, however, drew attention to Ukraine, and to the Ukrainian aspect of the disaster. Consequently, the Soviet authorities have emphasized not only Soviet, but specifically Ukrainian participation in the IAEA, and in developing nuclear energy in their own right. Unfortunately, such comments fall within the sphere of myth rather than reality, just as Ukraine, which like Belorussia, possesses a seat in the United Nations, has never yet voted against the Soviet Union on any question. Nevertheless, since 1987 marked the 30th anniversary of the IAEA, the occasion was used for celebrations in both Moscow and Kiev. The aim of such rejoicing was twofold: first, to demonstrate that the USSR has always participated actively in the organization; and second, to show that Ukraine has a long history of IAEA involvement. We will begin in reverse order.

Ukraine in the IAEA

The Ukrainian SSR has been a member of the IAEA since April 2, 1957, having joined the organization shortly after the Soviet Union did so on February 9 of the same year. According to the First Secretary of the Ministry of Foreign Affairs of the Ukrainian SSR, Borys Tarasyuk, the Ukrainian delegation played a significant part in the October 1956 conference in New York, at which the Constitution of the IAEA was elaborated and ratified.[2] In 1957, the organization had a membership of 26 countries, of which 3 consisted of Soviet members: the USSR, the Ukrainian SSR and the Belorussian SSR. Initially, the IAEA's main function was to restrict the expansion of nuclear weapons. It was considered that as several countries had the facilities to develop nuclear weapons, in addition to the 5 countries that already possessed them by the 1950s, there was a definite threat of a nuclear war that had to be averted. By the 1980s, the number capable of weapons production had risen from 5 to 20.[3]

At the IAEA, a system of controls was elaborated with the aim of ensuring that nuclear power plants and nuclear technology in those countries that did not have nuclear weapons would not be used for the purpose of producing them. By 1983, 850 nuclear plants were technically under the agency's control, and the prognosis at that time was that the number of nuclear power plants in the world would rise from 277 to 430 by 1990, accounting for 12% of the world's electricity supply. The IAEA has also maintained that nuclear power is one of the safest of all energy sources. It noted, for example, that if one looks at the number of fatal accidents that have occurred during the production of 10 gigawatts per hour of electricity,[4] then coal tops the list with 250, followed by oil with 200, wind energy at 70, solar energy with 60, hydroelectric power with 5, atomic energy with 1.5, and natural gas with 0.5.

This chapter will not analyze the organization of work at the IAEA in any detail. Suffice it to say that it has been a body with high prestige and a low budget. Its reputation has been enhanced by the Chernobyl disaster, but it is only fair to add that its directors have tried to promote nuclear power as the most viable of energy alternatives at every opportunity. It would be surprising had it been otherwise. Ukraine's active role in the organization coincides with the development of nuclear energy in the republic in the 1970s. Before this time, Ukraine's membership was symbolic rather than actual. Even in the 1970s, it is not clear whether the activities that have been noted with the benefit of hindsight ever amounted to serious participation. On the other hand, there is no doubt that in the 1980s, Ukraine became a leading center for Soviet nuclear research in addition to the expansion of nuclear power.

In 1967, the Ukrainian authorities established a Permanent Commission for Relations with the IAEA in the republic. It was composed of representatives of the Ukrainian State Planning Committee (*Gosplan*), the Academy of Sciences of the Ukrainian SSR, and the Ukrainian Ministry of Foreign Affairs. The avowed goals of the Commission were first, to expand scientific and technological contacts between the Ukrainian SSR and the IAEA, and second, to develop proposals on the use of international research in the introduction of atomic power into the national economy of Ukraine, which at that time had no nuclear power plants under construction.[5]

The Commission's main tasks, however, do not appear to have been very significant at first. It prepared brochures on "The Peaceful Atom of Ukraine" (1975) and "Atomic Energy in Ukraine" (1977), which was evidently distributed among delegates to IAEA conferences. In 1983, a new brochure was being written about the "successes in the peaceful use of atomic energy in the republic." Kiev was becoming a center for some international conferences on control over the nonexpansion of nuclear weapons among the "socialist nations." One such meeting took place in 1979.[6]

The republic also offered through the aegis of the IAEA some technical assistance to developing nations, making small donations—on average about 40,000 rubles annually—to the special fund that had been created within the IAEA. Specialists from 50 countries visited Ukraine between 1973 and 1983 (by 1987, the number of countries represented had increased to 60, encompassing about 800 specialists),[7] to carry out research on reactors, or the application of isotopes in medicine, biology and agriculture. For the most part, the visitors were taken to Ukrainian scientific research centers in Kiev, Odessa, Cherkassy and Krivyi Rih. The first official visit from an IAEA General Director to Ukraine seems to have been that of Z. Eklund in 1977, who arrived in Kiev (and also visited Yalta) with A.M. Petrosyants, Chairman of the USSR State Committee for the Utilization of Atomic Energy. The current General Director, Hans Blix, paid an official visit to Kiev in 1982, shortly after taking his post, to look at the Institute of Nuclear Research at the Ukrainian Academy of Sciences.

The republic reportedly developed a program known as *Energokompleks* (Energy Complex), which developed new techniques in energetics and environmental protection. It was reported also that in Ukraine was created the first "radioecological model," which described the migration of radionuclides in external surroundings, and upon completion would enable recommendations as to the most rational location for future nuclear power plants. As far as theoretics is concerned, Ukraine has an impressive reputation, but the practical application has not always been so thorough. Thus while Chernobyl may have been a logical place to site a nuclear power plant—the area had a plentiful supply of water, was sparsely populated, and a considerable distance from the major cities—the safety factor was reduced by the decision to build six 1,000-megawatt reactors at the same location. Moreover, the plans to build reactors close to major cities could hardly have been sanctioned through the study of a radioecological model.

Although the Ukraine-IAEA contacts have a long history, there has been nothing especially significant about the relationship. As noted above, Ukraine does not run its own nuclear power industry, and thus its importance in the nuclear program lies more within the realm of theory. When asked to speak about Ukrainian links with the Vienna-based organization, one of the Ukrainian delegates to the IAEA, Volodymyr Matviichuk, replied only that:

The Ukrainian SSR actively supported efforts to create the agency. It became one of its first members. And now a joint-committee on relations with the IAEA has been working in Ukraine for almost 20 years. This specialized organ carries out major work on the coordination of measures under the auspices of the IAEA program, and safeguards Ukraine's participation in the international exchange of information.[8]

In short, there was very little to say about a relationship that had lasted 30 years that could be applied to Ukraine as distinct from the Soviet Union. Nevertheless, in both Moscow and Kiev, the 30th anniversary of the IAEA, which was virtually ignored by the media of the Western world, was seen as an occasion for celebration in the Soviet Union. The question is why?

The answer is that there is a strong political dimension to the USSR's perception of the work of the IAEA that has very little to do with nuclear power per se. For example, to mark the 30th anniversary, Borys Shcherbyna, Chairman of the Government Commission set up to "eliminate the consequences" of the Chernobyl accident, sent a telegram to Hans Blix, which noted that the formation of the IAEA had been dictated by the need to solve the problems posed to mankind by atomic energy. However, this agreement had not gained a universal character and "negative tendencies" existed that threatened to undermine the regime of nonexpansion of nuclear weapons. What were these "negative tendencies"? The chief one was said to be the tendency of the United States to succumb to the pressures of "military-political interests" despite paying lip service to the nonexpansion on paper. In Moscow's view, responsible and decisive actions on the matter were required from the IAEA.[9]

The above is hardly of concern to the Chernobyl accident. But the Soviet Union was using the ties developed with the IAEA through Chernobyl to propagate party policy. Only after stating the above goals did Radio Moscow, through the medium of Deputy Minister of Atomic Power Engineering, Aleksandr Lapshin, outline the advances of the IAEA in securing the safe operation of nuclear power plants. Yet while it is true that the IAEA was founded originally to prevent the spread of nuclear weapons production—which has been somewhat akin to Canute trying to hold back the waves—its current prestige and reputation have been gained by the work undertaken following the Chernobyl disaster.

The anniversary celebrations were held simultaneously in Moscow and Kiev. In the latter city, a meeting was held on July 29, 1987, which was opened by the Ukrainian Deputy Minister of Foreign Affairs, A.M. Zlenko, and addressed by Deputy Chairman of the Ukrainian Council of Ministers, N.F. Nikolayev. The latter noted that in reality, the agency had become the first link in a system of measures and practical actions known today as the "international regime of the nonexpansion of nuclear weapons." Since this agreement had been signed in 1970, Nikolayev stated, the IAEA's role had grown in the effort to stop weapons expansion. Only later did he mention Chernobyl in the context of the Soviet Union's bid to create an "international regime for the safe development of nuclear power" with IAEA help.[10] One should perhaps not put too much emphasis on Nikolayev's speech, however, because only three weeks after making it, he was removed from the Ukrainian Council of Ministers after seven years of office, retired, and replaced by 50-year-old Viktor Hladush.[11]

Although we have argued that there was a strong political hue to the USSR's new emphasis on its commitment to IAEA policies and objectives, there was one major change in the sphere of nuclear power, which was that IAEA representatives were permitted to visit the Soviet Union's nuclear power plants, and especially Chernobyl, on several occasions, something that would not have been countenanced before the mid-1980s. Hans Blix was a visitor on three occasions between May 1986 and April 1987. On the first trip, he visited Chernobyl at the height of the post-accident crisis, when the Soviet authorities were by no means sure that their measures to quell the disaster would be successful. After the August 1986 Vienna meeting, it was expedient for the Soviets to allow further visits of Dr. Blix to ensure that the Chernobyl plant's modifications were adequate for its continuing operation.

The January 1987 visit may have been the most important. It was reported that the Soviet government extended an invitation for Dr. Blix and his colleagues to come to the USSR on January 13. He then toured the Chernobyl plant, noting the retrofittings that had been added to the older RBMK units 1 and 2 that were now both back in service. He met the highest-level officials in the nuclear industry and in spheres concerned with the cleanup and environmental situations: the Minister of Nuclear Power Engineering, Nikolai Lukonin; Petrosyants; Leonid Ilyin and Ukrainian Premier, Oleksandr Lyashko.[12] Dr. Blix was subsequently quoted by several Soviet sources as having given his approval to the changes made to the reactors, witness the following interview in a Ukrainian newspaper:

QUESTION: Dr. Blix, you were here last time when the situation was not quite as good as it is now. Then you flew in on a helicopter and now you're entering the station and visiting everywhere. Are you impressed with this progress?

ANSWER: Yes. It is almost eight months ago that we were here and they [had] just managed to get the situation under control. I'm very impressed with what has been achieved here in that period in bringing the two blocks into operation. Again, I'm impressed with all the modifications that have been undertaken at the station in the operative procedures to make it safer, and also to decontaminate and make this station safe for the operators. They are working under normal conditions inside the station now.[13]

From reading this interview, one would assume that the IAEA General Director had given unqualified approval to every measure undertaken by the Soviets, but as noted earlier, it was after this visit that the USSR abandoned longterm plans for the construction of the RBMK reactor. Thus it seems that in private conversations, Dr. Blix may have had some advice that differed from the publicly reported praise and support of the RBMK reactor.

The Soviet-IAEA cooperation in nuclear energy has generally and justifiably been perceived in a positive light. If the IAEA can assist in making Soviet (and other nuclear plants) safer, then only good can come of the relationship. At the same time, the IAEA is a United Nations organization, composed today of 113 member states. The Soviet Union has seen it in its interests to elaborate on its past and present role within the IAEA as an active participant and even to emphasize the individual part of Ukraine within this same organization. This sort of propagandizing tends to belie the more or less token participation of the USSR during the early years of the organization and gives an impression of a constant commitment to the agency's goals from the first. But there is something incongruous about the constancy shown by a country that has expanded widely its own nuclear weapons production while denying to other countries the right to develop their own (the role of the United States in a similar program notwithstanding).

Today, the Soviet Union is also one of the most aggressive proponents of nuclear power (some of the arguments currently in vogue in support of the industry are examined below) and is building up its own industry to a degree unprecedented in Soviet history, while the IAEA has been one of the organizations that has struggled to maintain the standing of nuclear energy in the world during the time of its most enduring crisis, namely Chernobyl. Thus, if only for a brief period, Soviet and IAEA interests have coincided. The most enduring evidence of this cooperation was the IAEA meeting in Vienna (discussed in the Introduction), and the conventions that ensued.

The IAEA Conventions of September 1986

The meeting of experts at the IAEA in Vienna in late August 1986 (described in the Introduction to this book) had been approved at a special meeting of the IAEA Board of Directors on May 21, 1986. At the same meeting, it was also proposed to hold post-meeting conventions to give binding status among member nations to the current IAEA regulations on the providing of information in cases of nuclear accidents and emergency aid on the part of neighboring states, known officially as the Convention on Early Notification of a Nuclear Accident and the Convention on Assistance in the Case of a Nuclear Accident or Radiological Emergency. The draft texts for these two conventions were drawn up between July 21 and August 15, 1986, and discussed at the IAEA General Conference that followed the post-accident review meeting and the two Conventions, on September 24-26.[14]

The first Convention noted that in the event of an accident at a nuclear reactor, waste facility, or during the transport and storage of nuclear fuels, nuclear wastes or radioisotopes, the "State party" must notify those states in the vicinity of the accident, either directly or through the IAEA, providing details about the precise time and location of the accident. The state in which the accident occurs must also provide any information that would result in the containment of such an accident's radiological consequences. Included in the details to be provided about the accident are the causes, the composition of the radioactive materials released, meteorological and hydrological conditions, the measures taken to combat the accident, and the longterm fallout that might be expected.[15]

The Convention concerning assistance in the event of a nuclear accident specified that either one or several states were to come to the aid of the "State party" in order to minimize the effects of an accident. If any country, whether or not the accident took place in that country, feels that it requires assistance from other states to lessen the impact of that accident, then it can either appeal to those states directly, or operate through the IAEA to request assistance. The member states should let the agency know of the equipment it possesses to combat such accidents, and the financial terms that might be drawn up for the use of such equipment. The role of the IAEA in the event of such a request is to allocate its own resources where necessary, pass on the request to other countries as required, and if asked to do so, to coordinate assistance "at the international level."[16]

In November 1986, when the Presidium of the USSR Supreme Soviet ratified the two Conventions, it was noted that neither one specified that states must report accidents at military installations. However, it was reported, the five major nuclear powers, including the Soviet Union, have all stated that they will report on those accidents also.[17] But what would happen in the case of disputes between states or between a state and the IAEA in the interpretation of the two Conventions. Article 11 of the first Convention was quite explicit on this point:

> If a dispute of this character between States Parties cannot be settled within one year from the request for consultation pursuant to paragraph 1, it shall, at the request of any party to such dispute, be submitted to arbitration or referred to the International Court of Justice for decision. Where a dispute is submitted to arbitration, if, within six months from the date of the request, the parties to the dispute are unable to agree on the organization of the arbitration, a party may request the President of the International Court of Justice or the Secretary-General of the United Nations to appoint one or more arbitrators. In cases of conflicting requests by the parties to the dispute, the request to the Secretary-General of the United Nations shall have priority.[18]

It was possible for a state to ratify the two Conventions without considering itself bound to the above, which was a curious loophole, but probably the only way in which to obtain the necessary ratification from the member countries. Thus if Chernobyl had occurred in 1987, for example, Sweden would have had a case for maintaining that the Soviet Union had not followed the rules for early notification—which is what actually happened in 1986 at which time the USSR, although a member of the IAEA, had not followed the IAEA's general guidelines on accident notification. But the Soviet Union, while agreeing in principle to all the stipulations of the Convention, would not have had to agree to the means of settling the dispute. And in fact, on November 14, 1986, TASS declared that the Soviet Union did not consider itself bound by the two sections of the Conventions that allowed international arbitration in the event of a disagreement. The USSR preferred to bar arbitration unless all sides were in agreement.

The loophole is indicative of the limited role played by the IAEA as the world leader in trying to ensure more safety in nuclear energy operations. In effect, the rules cited above can be broken in practice by any member state, since that state can veto any attempt at international arbitration in the case of a dispute. This is not to say that if all member states act in a responsible manner, the Conventions would not have validity and do much to lessen the impact and repercussions of an accident. But the loophole does exist, and the Soviet Union, whose actions were responsible for the Conventions in the first place, while acting quickly to sign and ratify them, was also not slow to reject Article 11 of the first Convention.

Having attended the Vienna meetings of August-September 1986, the Soviet Union then turned to its own nuclear industry, and re-examined in particular the safety of the RBMK-1000 and the RBMK-1500.

Changes to the RBMK

The changes made to the RBMK were determined by what the Soviets perceived as the causes of the original accident: a combination of human and design factors, with the former bearing the brunt of the blame. There had been a notable reluctance on the part of Soviet officials to criticize the design of the RBMK, even though a considerable amount of time was spent on making modifications, introducing technical retrofittings and devising new laws and technology for the control rods of this reactor. For example, one account stated that the post-accident analysis had not revealed any notable defects about the RBMK, its equipment or technology (which was patently untrue). The creators, the article continued, had failed to anticipate only one thing: human failure. Evgenii Larin, a Deputy Director of the Institute for the Operation of Atomic Energy Stations, also commented that although all Soviet RBMKs had been shut down in 1986 for retrofittings, the revisions had not revealed "one defect in the model or error in the plan for the creation of the atomic energy station." He also believed that it was man and not the machine that had caused the breakdown, which culminated in the disaster.[19]

The Soviet view is that the machine would have worked but for the errors of man. This reverence for the machine and for technological progress is something that is ingrained in the Soviet mentality, and even dates back to Lenin. One can compare the Western view that machines can and will go wrong, and the most important factor in building any nuclear power plant is foreseeing every possible type of accident, including all those that could possibly be caused by human error. Thus, although the Soviet interpretation of the accident is different from that of the West, there need not be any major disagreement over the sequence of events. The only argument is over the culpability for such a sequence. We will see more of the Soviet concept of the infallibility of technology below.

The most obvious drawback of the RBMK design is its positive void coefficient, the fact that at lower power levels, the reactor becomes unstable. To offset this difficulty, new laws on operating the reactor were introduced and some changes were made to the control rods themselves. A.M. Petrosyants confirmed that the accident had demonstrated the insufficient speed of response of the RBMK's accident prevention system. Before the accident, the safety rods had been located above the reactor's active zone, but were now being placed 0.7 meters into the active zone in order to facilitate a more rapid response.[20] By May 1987, it was stated, changes had been completed that would prevent a power surge occurring if rules of operation were again violated, and in addition, in a "scram situation," the rods could now be inserted into the reactor core in 10-12 seconds, as compared to 18-20 seconds required formerly.[21] But although the shutdown time had been reduced, it would still not be enough to prevent an explosion should a Chernobyl-type situation be recreated because the time required for the power surge that took place in April 1986 was only 4 seconds.

The Soviet report to the IAEA meeting that took place in Vienna in late September-early October 1987 indicated that the fixing of the rods into the active zone (the source states a depth of 1.2 meters rather than the 0.7 meters stated above) had not been successful because it

> ...led to a distortion of the vertical fields and made it necessary to reduce the power by 10 to 15%. At present the design of the rods has been changed: the connecting link between the rod and the "displacer" has been lengthened so that it is now possible to reduce vertical alignments of energy release.[22]

The minimum number of control rods during the reactor's operation was raised from the original 30 (see the Introduction) to 70-80.[23] At 30 rods, the reactor would now shut itself down automatically. To lower the positive void coefficient, the fuel enrichment of the reactor was to be raised from 2.0 to 2.4%. According to the September-October 1987 Soviet account, reactor tests on 146 fuel assemblies with an enrichment of 2.4% were performed at the oldest Soviet RBMK plant, Leningrad. In addition, a trip had been installed to prevent personnel from operating the RBMK-1000 below 700 megawatts thermal power. Another means of reducing the positive void coefficient was reportedly to place a limit on the amount of graphite in the pile and in the reactor, mainly "by shearing off the ribs of the graphite blocks."[24]

The human chain of command at the reactor was also amended in the light of the accident. In January 1987, Valerii Legasov, First Deputy Director of the Kurchatov Institute at the Soviet Academy of Sciences, said that a multi-step system of command had been developed for very important decisions, so that the key to emergency equipment was in the hands not only of the operator, but also of the "engineering hierarchy."[25] In other words, it would no longer be possible for an operator to violate operating rules by bypassing or switching off the reactor's safety mechanisms. Petrosyants noted that during planned upsurges of power and stoppages of the reactor, the presence of representatives of the State Committee to Ensure the Safe Operation of Atomic Energy Stations (*Gosatomenergonadzor*) "is essential," and that if a member of the Committee prohibited an operation from being carried out, then this decision was "definitive."[26]

At the time of the accident, it is clear that officials from the *Gosatomenergonadzor* were present at the Chernobyl-4 unit, but the above statement implies that their advice about the wisdom of carrying out the tripping experiment may have been ignored or overruled. After a reactor shutdown, for any reason, the reactor cannot be restarted, according to the new rules, unless the Chief Engineer of the nuclear plant, "or in extreme cases, his deputy," is present. According to the Chernobyl Director, Mikhail Umanets, anyone who allows the slightest deviation from regulations in future will quickly lose his position. He added that such rules were being applied not only at Chernobyl, but at every nuclear power plant in the Soviet Union.[27]

Another aim of the Soviets in the running of nuclear power plants has been to attain the maximum level of computerization of information about the radiation system and other factors. Academician V. Maslov has declared that as mankind clearly cannot get along without nuclear power plants, it is essential to ensure their total safety by running every possible accident variant through a mathematical model. Hitherto, he admitted, this had never been carried out in the Soviet Union, but Chernobyl had provided two lessons for Soviet scientists: first, that any deviations from "mathematical" precision can bring about a catastrophe; and second, that experimentation on a working reactor is inadmissible. In his view, full computerization would enable the study of the different variables of reactors, their diverse states and accident situations, including explosions, and their possible causes. Such a system would be expensive, he acknowledged, but the costs would be justified.[28]

Maslov's remarks indicated that such a system at Soviet nuclear power plants was in its infancy. Indeed, in mid-June 1987, Radio Kiev stated, in a confusing broadcast, that at Chernobyl nuclear power plant, an automatic "system of control" over the radiation system was being created "continually" and would eventually become the norm for other Soviet nuclear power plants. The goal was to ensure the automatic collection of data and their computerized analysis.[29] A meeting of the CC CPSU Politburo was convened to discuss this same question on July 30, 1987, and reportedly adopted a series of measures aimed at the establishment of a "control system on a high technical level."[30] The Soviet nuclear energy authorities were evidently convinced of the need for more advanced computerization of the industry, but by 1987, were still some way from completing research and application in this sphere. The accident had thus prompted new thinking in a number of areas.

Also under review was energy equipment, which had traditionally been of low quality. At a meeting of the CC CPSU in April 1987, Vladimir Dolgikh informed participants that:

> We must change our attitude toward questions of raising the technical level of energy equipment. It is not meeting today's demands of economy, reliability and endurance. The low quality of the equipment that is sent to energy units results in a great loss in the safeguarding of a continuous supply of energy and often leads to buildings with factory defects....

He singled out specifically the Ministries of Power and Atomic Power Engineering of the USSR, which he felt were not sufficiently close to the sphere of new research and technological achievements. Many electric stations, he added, were not looking after their equipment properly. It was being left out in the open and thus subjected to damage, and technological rules were violated in the use of such equipment.[31]

According to Dolgikh, the Soviet Union was lagging behind foreign models in terms of machine-building. Referring to the computerization question cited above, he commented that at "our electric stations we cannot get away from primitive recording devices and antiquated equipment," but that it was essential to computerize the operation of power units. Much of the equipment that was being produced for power plants traditionally came from the same factory. Dolgikh felt that the quality of the equipment would be improved if other factories produced competitive and rival models. Instead, the country's power stations were being impeded in their operation by rushwork at a single enterprise, which discredited "even the best of designs." Chernobyl, he concluded, "will long remain in the people's memory," but his implication was that the catastrophe had not led to a significant improvement in the quality of energy equipment.

Dolgikh's speech was an important critique of the operation of the Soviet nuclear industry. It revealed that although the Soviet authorities might have applied themselves effectively to the intricacies of the RBMK design—although even in this case they could not eliminate the fundamental flaw of that design—when it came to more basic questions such as computerization or the quality of equipment at nuclear power plants, Chernobyl had changed little.

According to an article by Andrei Pralnikov in *Moscow News*, the main problem both in the past and present was that all the building work was linked to a precise and tight schedule, and that the task was consequently carried out "Only to make a report. At any price." Following the completion of a poor-quality edifice, considerable time would then be spent on eliminating all the defects within the structure. The article observed that the staff at Soviet nuclear power plants was three times greater than that at foreign nuclear stations of similar size. The number of operators and engineers was the same, but the Soviet plants required an enormous number of maintenance workers. In explaining why this should be the case, Pralnikov offered a devastating criticism of the way equipment was being assembled at Soviet nuclear plants:

> [There is] practically no item of equipment that could be put into operation without "putting it into order"—that's how our industry works. And the thing "put into order" calls for keen attention later on. Even though every theoretical and practical worker knows very well that to make something work well at once is both cheaper and more beneficial. But alas, in the zone and at the station when the cleanup operation was underway, old mistakes were repeated.[32]

The main change that has been implemented in the area of energy equipment was initiated before Chernobyl, which is the unification of reactor and equipment design implemented at the Zaporizhzhya VVER-1000 nuclear plant in Ukraine, but the idea had been primarily to raise the tempo of reactor building rather than to ensure better quality of equipment. In this, it has been successful (see below), but the situation generally in the Soviet nuclear industry in the first 20 months after the accident, in the area of automization and assembly work, could only give rise to concern, if not alarm that the haphazard assembly measures of the past had not been improved to any significant degree.

Operator Training

The question of operator training loomed large after the Chernobyl disaster. It had been reported that the operator involved in the fatal experiment had no previous experience at nuclear power plants, and, moreover, that the accident itself was caused through gross violations of operating procedures by plant personnel. It followed that a thorough review of the plant operatives would follow. Following the restarting of Chernobyl-1 and 2 units, the plant's Shift Supervisor, A.G. Shadrin, was asked whether there was any difference in the "discipline, vigilance and responsibility" shown by the operators "then and now."

He responded that the authorities had taken a second look at many questions. The plant personnel were given a special retraining course and had to obtain their certificates over again. They were then put on standby operations at other related enterprises and were given a medical screening:

> A series of educational measures have been applied and worked out. The main form is individual attention to each plant operative at every level—from the chief specialist to the rank-and-file worker. Nuclear plant staffs have learned their recent lesson well: the number of those constructing a plant runs into the thousands, the number of those putting it into operation can number hundreds, but an accident can be caused by just one individual.[33]

Umanets was quoted as saying that the lesson learned from Chernobyl was that a drastic improvement was required in the training of personnel. How was this to be brought about? For one thing, each worker was to be subjected not only to a program of retraining, but also a psychological checkup. In addition, a full-sized simulator was being introduced into the training program for work on Soviet RBMKs.[34]

Before the Chernobyl accident, very little was known about the education and qualifications of operators at Soviet nuclear power plants. It had been assumed that the Soviet nuclear industry relied on simulators for the training of plant personnel. The simulator emulates critical situations that might arise during the work of a nuclear reactor. It can be used to train personnel both for work in the industry and for retraining sessions in the event of an operator's lengthy illness or other enforced absence from the nuclear plant. At the IAEA meeting in Vienna, the Soviets had in fact confirmed that they had simulators for the RBMK and VVER-type reactors, without going into any details. It was revealed subsequently that they did not possess any simulators for the RBMK, and had just one for the VVER, a somewhat outdated version at the Novovoronezh nuclear power plant.

In early April 1987, *Pravda* stated that the decisions of the CC CPSU issued after the Chernobyl accident had been ignored. There was still only one training simulator in the entire Soviet nuclear industry. In 1978, when the Soviets had first introduced the Novovoronezh simulator, it had been one of 20 in the world. But whereas the world now had 100 in operation, the Soviet figure had remained at one, even though it had been decided to build a second simulator at Novovoronezh in the late 1970s. More important, a resolution had been accepted at this same time to construct a simulator for the RBMK-1000. It had not been built. Hence, none of the operators at Chernobyl could have practised their craft on a simulator before starting work on the reactor itself. To the adjective "overconfident" which has been attributed to those conducting the experiment at Chernobyl-4, one should therefore add "undertrained." "How can we train operators with no training devices?" asked *Pravda*.[35]

Two weeks after the appearance of this article, Radio Kiev announced that a new simulator had been designed by Kiev scientists for use at the Chernobyl plant.[36] The news was broadcast on the international channel and may not have reached domestic listeners. In any case, it was an isolated announcement that remained uncorroborated by other sources. The radio had stated that the simulator at Chernobyl would be the first of several to be used in the Soviet nuclear industry. In May 1987, Nikolai Lukonin, Minister of Nuclear Power Engineering, stated that new training centers were soon to be opened at Novovoronezh and Smolensk (RBMK-1000) nuclear power plants.[37] Lukonin also presented a paper at the Fall 1987 IAEA meeting, in which he again stated that "training-simulator" centers were being created at Novovoronezh and Smolensk. He added that in 1988, 2,000 plant operatives would participate in a full course of special training and that considerable attention was being paid to the further improvement of simulators. They would eventually be installed at other nuclear power plants, but evidently, as far as the RBMK-1000 was concerned, Smolensk, rather than Chernobyl, was to receive the first simulator.[38]

In mid-May, Novosti announced the opening of a "center to train personnel" at the Smolensk nuclear power plant,[39] but this seems to have been premature. The new center opened only in September 1987. Lukonin also pointed out that the demands on the plant Directors had also increased. The Director should, he felt, not only be fully acquainted with the physics of the reactor, but he must have organizational skills and the ability to work with people. Many specialists were now being assembled at the newly formed Ministry of Atomic Power Engineering, he stated, and an experimental base was being developed within the Ministry.[40] While belated, such measures in the Soviet nuclear energy industry could only be welcomed.

The practice of sending new recruits in the industry immediately to work on reactors in active service had reportedly been ended. As far as the Chernobyl plant was concerned, the recruits were sent from their institute of training to further "study-training points" in Kiev, after which they spent another 3 months at a similar nuclear power plant before arriving at Chernobyl.[41] At the restarted Chernobyl-1 unit, most of the engineers came from Leningrad nuclear power plant, where, as noted, many tests were conducted to improve the safety of the RBMK-1000. They then had to undergo a period of 2 weeks' special preparation before being allowed to take up their new duties.[42] Lukonin maintained that the training for work at nuclear power plants had become so rigorous that it rivaled that of the Soviet cosmonauts.[43] In the view of the Soviet authorities, employees in nuclear power are much better trained than was the case before the Chernobyl accident.

It is important, however, to put matters into perspective. Most of the above announcements were made only one year after the disaster. Although a simulator may have been in place at Smolensk by September, it seems hardly likely that it was functioning properly. In his paper to the IAEA delivered at the meeting of September 28-October 2, 1987, Lukonin clearly implied that the training of personnel on simulators at Smolensk and other stations still lay in the future:

> In 1987 the training of the operating personnel in the USSR for units with VVER-1000 reactors will be expanded at a training-simulator center at the Novovoronezh nuclear power plant; the same thing *will be done* for operators of units with RBMK reactors at a training-simulator center at the Smolensk nuclear power plant.[44] [Author's italics.]

Yet the Chernobyl plant, at which the various transgressions of rules had been carried out, had already been restarted. If the Soviets were serious about the significance of training on simulators, why had they restarted one of their largest RBMK plants before the first simulator for work on the RBMK was in place? The answer appears to be that the training would take place without interfering with the nuclear energy program. But which was of more importance: the safety of the plants or the country's need of the electricity? Thus one might give a stamp of qualified approval to the various measures undertaken for the training of operators. They were badly needed, but there is also no doubt that the Soviet RBMKs in the 1980s, as throughout their history, were still being run by operators who had not experienced the benefits of a complete and effective training program.

The Soviet Pro-Nuclear Power Camp

Chernobyl naturally had a major psychological impact on the Soviet Union, just as it did on other countries of the world. Where the USSR has differed from elsewhere, however, is in the solid entrenchment of the pro-nuclear energy "camp" within the country's power structure. The industry is highly centralized and has become an integral part of the Soviet energy program, upon which depend Mikhail Gorbachev's plans for the "technological acceleration" of the Soviet Union into the 21st century. Moreover, in contrast to Western countries, in the USSR there does not appear to be a strong anti-nuclear power element among the scientific elite. In fact, whereas some politicians, literary figures and economists have opposed nuclear energy for a variety of reasons—some of which will be explored below—Soviet scientists have always come to its defense. To date, they have always had in the somewhat one-sided debate the powerful backing of the Soviet party hierarchy, which has invested in a huge program for nuclear power expansion and does not intend to change course in midstream.

One of the first public commentaries on the question appeared in *Pravda* in October 1986. The writer, A. Pokrovsky, stated that because of several accidents at nuclear power plants in various countries, general opinion on nuclear energy was now divided. Its opponents had managed to convince the governments of Austria and Denmark not to build stations on their territories. These same people, it was claimed, also tried to influence the post-accident review meeting in Vienna. Many honest people, said Pokrovsky, were 100% opposed to nuclear power, while other no less honest people were 100% in favor of it. "So who is right?" In his view, the Vienna meeting had provided convincing proof that nuclear power remained the best means of electricity generation for the future.[45]

The first question to be asked, wrote Pokrovsky, is can the world manage with a lower demand for electricity than at present? The answer, he said, was that despite some economizing in industrial expenditure of electricity, it was "unthinkable" that the country could meet social and economic demands without raising electricity production in the future. The dimly lit Moscow streets and cars driven with only parking lights on at night are testimony to his conclusion. His second and related question was: "at what expense shall we obtain it?" There was indeed a rather impressive list of alternative sources of electricity: coal, oil, gas, hydroresources, biomass, wind and solar. However, for the long-term future only three of the above—coal, oil and gas—could be counted upon along with nuclear power. Of these four, wrote Pokrovsky, the nuclear alternative was the least expensive and the ecologically cleanest source.

The Chernobyl disaster appeared to have obscured the record of the Soviet nuclear industry's otherwise excellent record of 4,000 reactor years of stable exploitation, he continued, thus "it is essential to restore the positive service record of nuclear energy." Was this comment accurate? Certainly Pokrovsky interpreted the Soviet industry in the best possible light. Could it not also have been stated, for example, that Chernobyl was the first *officially reported* accident in 4,000 reactor years? Clearly there were others, some of which are even now coming to light, such as the Rovno case mentioned earlier.

Pokrovsky felt that many people were simply confused by contradictory reports from the Chernobyl accident. He noted, with some justice, the diverse definitions of what constituted dangerous radiation levels in some food products:

> Aren't people really confused by the various approaches to the definition of a dangerous level of radiation? In Great Britain and Sweden, for example, milk is considered unsuitable for consumption if it contains more than 2,000 becquerels [per liter] of iodine-131. In Poland, the level is 1,000 becquerels, in Hungary 500, Austria 370, and in Hessen, West Germany, 20.

For the future, said Pokrovsky, what mattered in nuclear energy was international cooperation in applying safety techniques, particularly through the IAEA.

The same question was taken up by an expert, Valerii Legasov, in a series of interviews in the Soviet press throughout 1987. It should be noted at the outset, that Legasov is an outspoken proponent of nuclear energy. Legasov spoke from three basic suppositions: first, that nuclear power is the most economical of all energy alternatives; second, that it is the safest alternative, and that the accidents that have occurred stem from the human element in the man-machine relationship; and third, from the ecological perspective, that it is the cleanest of all the sources of energy.

He noted, for example, that at the end of 1986, the Soviet Union depended on nuclear power for only about 11% of its electricity production. This could be compared with the much higher percentages in other countries: 70% in France, 60% in Belgium, 43% in Sweden, 40% in Switzerland and Finland, and 31% in West Germany. But the question, in his view, was whether it was possible today to refrain completely from the use of nuclear power and return to a dependence on coal, oil, gas, oil products and hydroresources. His answer was that raw material resources would not permit such a dependence in the longterm.[46]

Legasov's view was that nuclear power "is a historically normal pattern of technological progress." This was not to say, he informed, that other energy alternatives were not being explored in the Soviet Union. At Kamchatka, a geothermal station was being developed; there was an experimental tidal-station at Kislogubskaya on the White Sea; and the country's first solar energy station was under construction in the Crimea. There were also designs for hybrid solar-thermal energy stations. But none of these could, in Legasov's opinion, guarantee large-scale production of electricity in the foreseeable future. In fact, he pointed out, in the future there are likely to be major political conflicts over fuel-energy resources throughout the world. Even resources of organic fuel could not last longer than the year 2100, he stated. As a result, nuclear power could play a stabilizing role in these conflicts.[47] In short, technological progress anywhere required a continuing and increasing supply of electricity.

But what about accidents such as Chernobyl? Legasov's views on the matter were very specific:

> The accident at Chernobyl, at Three Mile Island, and other tragic events not connected with the peaceful atom, for example, the explosion of the "Challenger" spaceship, the explosion in Bhopal, India, catastrophes at sea and on the railroad, have demonstrated to us that the problem of the interrelationship between people and the machine has still not been fully resolved and demands our tireless attention. The enemy is not technology itself, but our incompetence, our irresponsibility in dealing with it.[48]

The solution to the man-machine relationship was to develop a system in the nuclear industry in which the "strong points" of the machine could compensate for the shortcomings of operators, and machines with defects could be dealt with by staff who have the knowledge to deal with them. There was no greater risk of an accident at nuclear plants than at any other industrial facility, said Legasov.[49] But was it not possible, as one of his interviewers asked him, to build nuclear power plants further away from the major population centers of the European part of the country, in remote areas where they did not present such a danger to the surrounding population in the event of a release of radioactive materials?

Such an option had been considered and rejected, responded Legasov. For one thing, it would be uneconomical to locate plants so far from the consumers of the electricity. Moreover, there could not be a remote area as far as a nuclear power plant was concerned. A city would inevitably crop up near the plant site for the building workers, operatives and their families. Besides, areas that are remote today, he added, may be at the heart of a large population in the future as the Soviet Far East and Siberia are developed. Although nuclear plants should not be located too close to major population centers, it was justified to build them in the areas in which the main part of the population lived and in which the main industrial centers were located.[50]

As for the alternatives, both Legasov and Petrosyants pointed out in detail the ecological problems acquainted with the use of coal to fuel thermal electric stations. They noted that in the world today, from the coal, oil and natural gas burned every year throughout the world are emitted into the atmosphere about 200-250 million tons of ash, and about 60 million tons of sulfurous anhydrides. By the year 2000, they estimated, such emissions will have grown respectively to 1 billion and 300 million tons (Petrosyants) or to 1.5 billion and 400 million tons (Legasov). Petrosyants stated that nuclear plants provide more protection for the environment because their fuel-energy capacity is 10 million times higher than that of organic fuel. The expansion of nuclear plants in the Soviet Union would lead to a reduction in the scope of mining work, its influence on the environment and also a reduction in expenses on transportation. The annual demand for fuel at a 1,000-megawatt nuclear power plant, for example, is about 140 tons of natural uranium, whereas according to Petrosyants, for a thermal power station of similar capacity, 2 million tons of coal annually are required for its operation.[51]

The Soviet scientists received strong backing in their arguments in support of nuclear energy from the General Director of the IAEA, Hans Blix. Blix believes that any form of energy has its defects, and that the basic choice today for all countries is one between nuclear power and coal. He has maintained that the pendulum of public opinion is swinging toward nuclear energy, which in 1987-88 produced 15% of the world's electricity.[52] There have been many arguments against greater dependence on coal for energy requirements in the USSR, not least the frequent accidents that have occurred in the industry over the past few years.[53] Petrosyants stated that coal simply could not keep up with the rising demand for energy:

> If you accept the annual growth of [energy] demand per capita to be 3% and that the population in the year 2000 [worldwide] will be at least 5 billion (according to the United Nations, 6-6.5 billion), then by 2050, specialists believe, the possible expenditure of energy will exceed more than twice the energy equivalent of all the currently known supplies of coal.[54]

Generally, therefore, Soviet scientists have supported the development of nuclear energy in the USSR. They admit that the industry contains an element of danger, but consider that this is simply one of the drawbacks of technological progress. The answer, in their view, is to ensure that nuclear power is made as safe as possible, and to establish an equilibrium between man and machine rather than to deny technical progress, "as foreign hotheads propose."[55] Although nuclear power has its drawbacks, it is, in their view, the best of all the alternatives, and its safety record is among the best of all the energy-producing industries, whereas coal, the obvious alternative, has the worst accident record in the group.

Finally, the number of casualties caused by Chernobyl was and will be, they believe, relatively small. Petrosyants referred to a long list of more serious accidents to have occurred in other industries that did not receive the same sort of publicity given to Chernobyl. It included a 1974 explosion of ammonium nitrate in a U.S. city that reportedly resulted in 576 deaths; a railway accident near Toronto in 1979, concerning a train carrying oil fuel, petrochemical products and chlorine that forced an evacuation of 250,000 people from the area; and the December 3, 1984 tragedy at the Union Carbide chemical plant in Bhopal, India, which led to 2,500 deaths and 100,000 injuries.[56] There is no question that Chernobyl caused a greater media sensation than these accidents, although unlike them, it was a disaster that had a world as opposed to a relatively localized impact on the surrounding population. Petrosyants was also correct in stating that nuclear power receives more publicity than other energy industries, especially when something goes wrong. The alarming accident rate in the coal industry, for example, attracts virtually no media interest. On the other hand, there had been a growing dissatisfaction in the Soviet Union with the way in which the nuclear industry is being developed and the apparent failure of the authorities to draw appropriate lessons from Chernobyl.

Opposition to Nuclear Power

Because of the official support for the nuclear power industry in the USSR, opposition to the industry—other than individual letters from anxious readers—was somewhat slow to manifest itself publicly, in contrast to some of the Soviet Union's neighbors. The first evidence of opposition in Soviet Armenia, however, preceded Chernobyl and expanded as a result of the accident. Following a protest against "environmental hazards" in the republic in an open letter from Armenians to Mikhail Gorbachev in March 1986, the Armenian authorities began to turn against nuclear power in the republic.

Thus in November 1986, Garen Dallakian, the Chairman of the Armenian Committee for Cultural Relations with Armenians Abroad, visited the Armenian community of Switzerland, and stated that the Armenian nuclear plant at Metzamor, about 20 kilometers from the Turkish border, would be out of commission by 1990, at which time the republic would use more conventional means of producing electricity. He cited the lack of safety features at the plant, and the fact that an accident had occurred there in 1982 as the main reasons behind this change of policy.[57] In January 1987, during an interview with a Soviet newspaper, Valerii Legasov was asked about the reports of the shutdown of the Armenian VVER-440 reactor because of "troubles" at the station. His response was to confirm the shutdown, but to claim that it had occurred because the Armenian nuclear plant was an old installation that now required technical modifications to bring it up to standard. These modifications, he added, were being made not in the light of Chernobyl, but rather as a result of the Three Mile Island accident report, which he and his colleagues had read in 1982.[58]

However, matters were more serious than Legasov indicated. Not only had the Armenian plant been shut down, but, according to the newspaper *Sovetskaya Armeniya* of March 17, 1987, work on building new reactors at Metzamor had also been brought to a halt. This fact had been revealed by the First Party Secretary of Armenia, Karen Demirchyan, at an Armenian Central Committee Plenum on March 14. Demirchyan confirmed that the Armenians would in the future become more reliant on hydroelectric stations and thermal electric stations which used natural gas.

The Armenian reactor in service had reportedly been leaking radiation, although this was denied by the councillor of the Soviet Embassy in Ankara, Anatolii Kadirov.[59]

In April 1987, even Petrosyants admitted that the Soviet authorities were "encountering certain difficulties in finding sites for nuclear power stations" because people were worried about nuclear energy.[60] The broad extent of these fears was suggested by Yurii Bublyk, who runs the Laboratory of the all-Union Scientific Research Institute of Fuel-Energy Problems at the USSR Gosplan. Where should nuclear power plants be built?, he asked:

> After the Chernobyl accident, the opinion of the majority of people was the same—further away from the heavily populated regions—in the deserts, tundra or mountains.[61]

But the above represented concern rather than opposition. The first serious protest in the Soviet Union after the Armenian affair occurred, not surprisingly, in the area of heaviest concentration of nuclear power in the country, the Ukrainian SSR.

The degree of cynicism among Ukrainians regarding official commentaries and analyses of Chernobyl can be gauged from the account in *Moscow News*, of a Ukrainian author who appeared on Moscow Television:

> We are being assured that the situation is absolutely normal and well-nigh better than it was before the accident at unit 4 of the Chernobyl AES. Is there really no room for improvement? Let's blow up one more unit to make the situation really splendid![62]

By August 1987, the first official protest from Ukrainians appeared in the Kiev press, in the form of a letter by several Ukrainian writers from Cherkassy Oblast demanding that construction work be halted at a new nuclear plant near Chyhyryn on the Kremenchuk Reservoir on both ecological and historical grounds. Among the signatories was F. Morhun, First Party Sceretary of Poltava Oblast. The letter began by stating that Chernobyl remained as a pain in the hearts of millions that had demonstrated that the peaceful atom can become a source of great evil if treated carelessly. At the present time, the writers continued, the troubles of Chyhyryn on the Dnieper, "beat more and more powerfully in our hearts."[63]

Chyhyryn was one of the most recent developments in the Soviet

nuclear energy program at that time. The proposed construction of the
new nuclear plant was revealed by *Izvestiya* on June 20, 1985. The plant
was to be a water-pressurized (VVER) reactor station, and according
to the late Theodore Shabad, would most likely have been between 4,000
and 6,000 megawatts in capacity when completed. Its location was the
site of an abandoned central electric station on the southern bank of the
Kremenchuk Reservoir.[64] The first project had entailed the resettlement
of the residents of the villages of Stetsivka, Vitove and Hushchivka, which
are located about 12-16 kilometers from Chyhyryn itself.

The article noted that these three villages had been resettled three times
in the past two decades, although in the case of Vitove and Hushchivka,
the period of turmoil was even longer. In 1958, they were transferred
to Chyhyryn Raion from the Kremenchuk Reservoir territory, in connection with the building of the Kremenchuk hydroelectric station.[65] To
compensate for the losses incurred on the first project, the Soviet
authorities decided to use the same location for a second, equally
ambitious project.

The writers noted that in 1969, it was announced in the press that
by the end of the 9th Five-Year Plan (1971-75), one of the largest power
plants in the world would be in operation at Chyhyryn, fed on coal, with
a capacity of 4,800 megawatts. Within a short time, however, it "became
obvious" that for such a capacity of power, there would not be enough
coal available, and therefore the proposed capacity was reduced to 3,200
megawatts. When this also was discovered to be impractical, the
designers, "not moving away from Chyhyryn, this sacred corner of nature
and Ukrainian history," changed the profile of the station again, this
time to one fueled with black mineral oil (*mazut*), with a capacity of 1,600
megawatts. But none of this type of oil was to be found in the pinewoods
and forest thickets of the area, and therefore the oil had to be fed by
pipeline from Kremenchuk. After the evacuation of the villages adjacent
to the station, the assembling of thousands of construction workers who
had already begun building work, and the allotment of 50 million rubles
for the first stage of the project, it was "suddenly revealed that oil cannot
be piped from Kremenchuk during the winter."[66]

As a result of this fiasco, a court case was held and the first Director
of the unbuilt electric station was put on trial, along with the Chief
Accountant. The project was abandoned. But although there were many
potential "buyers" for the site, "the energy workers clung stubbornly to
their captured base." Then, continued the letter, with bitter sarcasm,
"from the bureaucratic jungle of the department" came an even more
grandiose idea: to construct a nuclear power plant on the Dnieper. Some

of the more "woolly-headed" among the planners even wanted to ensure plentiful supplies of cooling water by building the station on an alluvial island in the middle of the river. But finally it was decided to build the plant on the bank of the Dnieper instead.

Although the design of the Chyhyryn nuclear power plant had yet to be confirmed, construction enterprises from the all-Union association "Union Atomic Energy Construction" (*Soyuzatomenerhobud*) had arrived in the summer of 1985 and had begun to work "feverishly." The article described a savage attack on the natural environment that resulted: "wide openings are being cut out as though they were working in a taiga," a boiler house had been started up and a concrete factory was preparing to begin operations. Some 3.5 hectares of pine forest had been hacked down. The people of Chyhyryn Raion were evidently so concerned about the new project that they had asked the group of writers to come and see for themselves what was happening.

The writers stated that they soon were made aware of the reasons for the anxiety of the population:

> In observing such negligence and mercilessness toward nature here every day for the past 17 years, and, what is worse, irresponsible design on the part of energy specialists, people are seriously perturbed about the future health of their land, the future health of the Dnipro. People are amazed and angered by the narrowminded unpatriotic approach to such a serious issue. It is one thing when the place, although contrary to common sense, is selected so far from the fuel for the power station, but quite another thing when here, on this little heel, in such a densely populated raion, on the bank of the Dnipro, the main drinking water supply of Ukraine, they want to put an atomic power plant.

"Is it possible," asked the writers, "that the Chornobyl tragedy taught us nothing?" They perceived the station as a distinct threat to the environment of the area, and especially to the cities located "downstream," such as Dnipropetrovsk, Zaporizhzhya, Dniprodzerzhynsk, Kremenchuk and Svitlovodsk, and the "millions in the Donbas" who drink the Dnieper water. In their view, the Dnieper had already been overworked as a source of energy. In the Chyhyryn Raion, it was harnessed for the Kremenchuk hydroelectric station; in the Cherkassy area, for Kaniv hydroelectric station; in the Kiev region, for Kiev hydroelectric station; and further down the river, for the Dnieper and Kakhivka hydroelectric stations.

However, the Cherkassy writers were also opposed to the Chyhyryn nuclear plant on historical grounds:

Let us be realistic. The building of an atomic energy station will transform Chyhyryn and Chyhyryn Raion into a zone closed to tourists. Even now, one of the most charming little nooks, the sacred, preserved, famous Kholodnyi Yar, rich in the history of our Motherland, has been practically fenced off from the world.

The writers provided examples of numerous historical monuments that would be endangered by the future nuclear plant. They noted that Chyhyryn itself was the historical capital of the Ukrainian Hetman state, at the time of the "national independence war of the Ukrainian people under the leadership of Bohdan Khmelnytsky, 1648-1654." At Chyhyryn, they noted, was born the "concept" of the union of Ukraine with Russia, and the harmony of the three Slavic nations—Ukraine, Russia and Belorussia. Thus they appealed to the patriotic instincts both of Ukrainians, and of the Soviet government in Moscow in the form of the Russian-dominated Politburo (at that time, the Georgian Shevardnadze was the only non-Slav in the CC CPSU Politburo, and with the exception of the Ukrainian Party chief, Shcherbytsky, and the Belorussian Party chief, Nikolai Slyunkov—a Candidate Member—all were Russians). Moscow has in fact traditionally celebrated the anniversary of the Ukrainian-Russian "union" (1654, highlighted in 1954 and 1979) with great pomp and ceremony.

So can we really shut this off from the world? Is this really our ideological narrowmindedness, dictated to us by narrowminded current interests?

At Kholodnyi Yar, there had occurred the "national revolt" of the late 18th century, led by Maksym Zaliznyak and Ivan Honta, but there were also more recent monuments. The writers, appealing to the feelings of the authorities, also stressed the monuments to the Great Patriotic War:

Here towers a monument to the partisans of the Great Patriotic War, and a monument, the Ukrainian Cottage, to the residents of the burned down country village Budy, who were tortured by Fascist brutes. Can we honestly deprive tourists, especially children, of these national shrines of our historical memory?

The letter stated that the writers were expressing the views of everyone with whom they had become acquainted at Chyhyryn. It maintained that the writers were not anti-nuclear, but that the location of Chyhyryn was disastrous for the area. But since the plant was still at the preparatory stage, "it is still not too late to halt construction." To continue, however, would do "irreparable harm" to nature. They proposed something innocuous in its stead, such as a fruit or vegetable-processing factory.

The letter to *Literaturna Ukraina* has been discussed here in detail because it was an unusually significant event. It was outspoken and angry, and yet was placed in a position of prominence in the newspaper at the top of page 2 in the form of an article rather than a letter, which suggests that the viewpoints expressed were implicitly endorsed by the editors. But on the other hand, it may have been a sign of the emerging openness on issues such as nuclear power, a demonstration of the fact that, as Andrei Pralnikov remarked, the "tendency to silence persistent opponents is gradually waning." The nuclear industry was no longer the preserve of Moscow-based scientists alone, as we have already witnessed from the public discussions in Kiev about the fate of Chernobyl-5 and Chernobyl-6.

The Ukrainian writer Oles Honchar followed up the protests of the Cherkassy compatriots at a speech during the All-Union Creative Conference in Leningrad on October 1, 1987. Honchar felt that ample warning of the April 1986 accident at Chernobyl had been provided by Kovalevska's article in *Literaturna Ukraina* one month before the event, but had not been heeded. Had it been read and had glasnost been in operation at the station then, in his view, the Chernobyl disaster might never have occurred. Instead, nuclear power plants were sprouting up all over Ukraine: at Rovno, Khmelnytsky, Zaporizhzhya, South Ukraine, and now Chyhyryn, which, he felt, was about to destroy a valuable historical area:

> There is a notion to build a plant also in the uplands of the Desna [River]. Who is to say that in each of these nuclear stations, built or planned, there is not a potential Chornobyl?[67]

However, in Honchar's view, "gigantomania" was prevailing throughout the Soviet Union, along with the view that "science demands casualties." His appeal was to the people who were being overlooked in the decision-making, but he was also dissatisfied with the post-Chernobyl analysis of Soviet scientists. In his opinion, the only way to discuss the issue properly would be to hold an international "Chornobyl forum." His speech was evidence that members of the Ukrainian Writers' Congress had started to constitute a more-or-less unofficial opposition where the Ukrainian nuclear power industry was concerned.

Traditionally, the Soviet scientific elite had not heeded such protests, and even today, in the era of "glasnost and perestroika," scientists are often disdainful of complaints against nuclear power development, especially from the sphere of literature. Asked about the Ukrainian protests, during a press conference of the Soviet Academy of Sciences in Moscow in late October 1987, Evgenii Velikhov, Vice-President of the Academy and a key figure in the efforts to quell the burning Chernobyl-4 reactor, responded that he was a scientist, and as such, "I cannot concern myself with the opinions of writers."[68] This was hardly an appropriate answer to the questions now being raised by many Ukrainians.

The first significant brake to be applied to the massive expansion of the Ukrainian nuclear industry was, however, applied almost immediately after Honchar's speech in Leningrad. Later in this same month, F.S. Temirov, the Director of the all-Union Atomic Energy Construction Planning Institute (*Atomenergostroiproekt*), made a surprising announcement during a review of current problems of atomic energy in the Soviet Union. Discussing the development of nuclear power and heating stations, which are to generate electricity in addition to heating major cities, Temirov stated that these stations were being built in Minsk, Odessa and Kharkiv, but that the plan to build one in the western part of the city of Kiev had been scrapped. Why? He stated that after Chernobyl, there had arisen a need to create higher barriers of safety, and therefore another thermal electric station would be built instead.[69]

The decision contradicts the above statements about the relative safety of nuclear power plants. Evidently they were not as safe as the Soviet authorities had made out. On the other hand, if the Kiev nuclear power and heating station has been abandoned for safety reasons, then why was construction going ahead on the Minsk station, which is located only 2 kilometers from the city boundary? The most likely answer is that Kiev was scrapped not for safety reasons, but rather as a consequence first of the fears expressed by the population after the Chernobyl disaster only 130 kilometers to the north, and second, as a result of the recent publicized protests against Chyhyryn and the development of nuclear power in Ukraine generally. Gradually, the fears of the public, in an era of glasnost, were having an impact on the nuclear power program despite reassuring comments from scientists.

In late November 1987, Valerii Legasov revealed that public opposition was seriously hindering work on both the Odessa and Minsk nuclear power and heating plants. As for Chyhyryn, one report notes that the Ukrainian Council of Ministers held a meeting on November 19 at which it was decided to discontinue the project. The source cited a telephone conversation with the head of the Writers' Union for the Cherkassy Oblast, "Nikolai Negoda."[70] It has not been possible to corroborate this statement. In fact, there are signs that, although Chyhyryn was now apparently fair game for any critic, work on the building of the station was continuing. Thus early in January 1988, in an article in *Sovetskaya kultura* that called for the punishment of those responsible for building Chernobyl, another Soviet writer, Evgenii Dudar, wondered whether

Perhaps these very same people from "Soyuzatomenergo" [Union of Atomic Energy] are today, despite the outraged voices of the public, building the Chigirin [Chyhyryn] nuclear power station?[71]

A day earlier, an article in *Molod Ukrainy* had made reference to the continuing debate about the future of the Chyhyryn station, although it made it quite plain that the proposed site of the station had been a grievous mistake.[72] On the basis of two Soviet sources therefore, there appears to be some doubt about whether the November 1988 meeting actually halted building work at Chyhyryn.

However, following the news about the commercial operation of two new Ukrainian reactors (in addition to an old one, Chernobyl-3) in December 1987, Zaporizhzhya-4 and Khmelnytsky-1, *Pravda* announced on January 21, 1988, that because of public opposition, plans to build the nuclear plant in the Krasnodar region, already well advanced, had been abandoned. A fuller explanation of this surprising event (hitherto few Western scholars were aware of the existence of a station at Krasnodar) was provided a week later by the newspaper *Komsomolskaya pravda*.

The article, which was written by V. Umnov, noted that the Chairman of the regional government in Krasnodar had manifested his opposition to the Krasnodar station on television, thereby changing the balance of public opinion on the topic. The newspaper pointed out that the Ministry of Atomic Energy of the USSR was now receiving letters expressing concern about nuclear power from Ukraine, Belorussia and from the entire country. The 20 "or so" existing stations and all the nuclear plants under construction were said to be bitterly opposed by local residents.[73]

For the first time in the fierce nuclear power debate that had emerged in the Soviet Union, there appeared to be some doubt about which side was winning. Gradually the opponents of nuclear power had begun to hold the upper hand. Within a matter of two months, the ambitious Soviet nuclear energy program, and especially its Ukrainian component, was dealt some devastating blows.

In January 1988, *Moscow News* featured a debate on the issue entitled "One Problem—Two Views," in which the arguments for and against nuclear power were outlined. Writing in favor of the industry was Dmitry Volfberg, Executive Secretary of the USSR State Committee for Science and Technology, whose arguments followed closely those of Legasov outlined above. Opposing nuclear energy was the economist Igor Reshetnikov from the Institute of World Economics and International Relations.[74]

Reshetnikov noted first that nuclear power had been curtailed drastically in the United States, where the building of more than 100 reactors had been cancelled. Over the past 12-14 years, he observed, the anticipated nuclear plant capacity in the West had fallen by 10-15 times. Why, he asked, in the face of foreign contraction had the Soviet industry begun a program of acceleration? The answer, he maintained, was that certain cliques of specialists—"clan interests of separate groups of scientists"—were providing partisan reports in favor of the industry.

Moreover, he believed, the program of development as set forward, was impossible to fulfil. The 1986-90 Five-Year Plan foresaw the commissioning of 41 million kilowatts (41,000 megawatts) of new capacities, but only half this amount was actually ready for the first phase of construction in 1986. It had long been realized, Reshetnikov continued, that these figures were simply unrealistic because there was no "corresponding reserve" for such a building program, nor had the necessary investments been planned:

> Incidentally, during the past five-year period 15.5 million kw were put into commission instead of the planned 24-25 million. It's all right! The decades of vicious practice in annually adjusting plan assignments have induced responsible officials and whole organizations to taking [sic!] a frivolous attitude towards various kinds of figures in long-term and even average-term planning.

Economically, stated Reshetnikov, it was inadvisable to build expensive nuclear plants. The arguments regarding emphasis on coal may have validity, but in his view there was another alternative for the immediate future, which was "the intensive modernization and replacement of obsolete capacities" at traditional electric power stations. The country could meet its electricity needs by raising the quality of electric power and by eliminating the "strained situation" at other power plants. Energy efficiency, he concluded, was one means of research for "new technical solutions."

In a mid-February issue of the same newspaper, the debate was continued. Among those who opposed the existing expansion program was Andrei Sakharov, the former dissident scientist. Sakharov maintained that accidents could not be predicted and that the measures that had been applied to improve safety levels at existing Soviet reactors were insufficient. Instead, he believed, the current work on building reactors should be stopped immediately and prohibited. Work should begin on a new program to build reactors underground where they would not only be safer to operate, but would also be provided with safeguards against acts of terrorism and a conventional war.[75]

Shortly before Sakharov's article appeared in *Moscow News*, thirteen Ukrainian scientists, including four academicians and three corresponding members of the Ukrainian Academy of Sciences, had published a lengthy polemic against the Ukrainian nuclear power program, which opposed nuclear plant construction on ecological, economical, agricultural and geological grounds. Among the scientists was Academician A.M. Grodzinsky, who had spoken out against the future development of the Chernobyl plant at the March 1987 meeting of scientists in Kiev. The article took the form of a direct attack on the Ministry of Atomic Power of the USSR, which it accused of acting without consulting the wishes of the people and which, in their view, was ripe for radical reconstruction.[76]

In particular, the scientists focused on the expansion of the Rovno, Khmelnytsky and South Ukrainian stations. It noted that a discussion had been held in the republic about the proposed expansion of these stations on August 25, 1987, along the lines of the one held about the advisability of pushing ahead with the third stage of the Chernobyl plant (see Chapter 6). In the original plans for the development of the three stations, only the completion of those energy units then under construction had been elaborated. It appeared, however, that the Ministry of Atomic Power had maintained that the Ukrainian republic required urgently a further 6,000 megawatts of electric power capacity, which it proposed to acquire by expanding the three stations above their scheduled sizes of 4,000 megawatts (in the cases of Khmelnytsky and South Ukraine) and 2,800 megawatts (in the case of Rovno). That the proposal, in the view of the scientists, was ill thought-out and unpopular, was made abundantly clear:

Nothing affects [the Ministry of Atomic Power's] efforts over and over again to push forward a decree that runs contrary to public opinion, and without calculating all the necessary components of a problem, explaining its tasks by whatever statement happens to be convenient at the time.

To build nuclear plants over 4,000 megawatts in size, wrote the scientists, ran contrary to the ecological appraisal of the plans for nuclear power development, which had been provided by various institutes of the Academies of Sciences of the USSR and Ukrainian SSR. The Ministry of Atomic Power's view was that the existing stations could be expanded more easily than new stations could be constructed. The infrastructure of settlement already existed and the local residents liked to put down their roots. Yet where was the guarantee, they asked, that in five years' time, the Ministry would not demand the further expansion of capacities at the three stations to 8, 10 or even 12,000 megawatts?

The Ukrainian nuclear power program was then attacked on agricultural grounds:

> The concentration of huge capacities in densely populated raions, the cutting off of water and fertile land, and social and ecological factors— make up a very important, thorny problem that worries scholars and specialists of the republic. The territory of the Ukrainian SSR has the highest level of economic usage in the country. In an area of 603,000 square kilometers (or 3% of the total area of the country) is produced today over 20% of the Soviet GNP, including over 25% of gross agricultural production. Ukraine also accounts for 20% of all the grain collected, 60% of industrial sugar beets, 45% of sunflower seeds, about 25% of meat, milk and potatoes, and one-third of all the vegetables grown in the country.

In addition, the scientists pointed out, the republic is also the vacation center of the Soviet Union, providing relaxation and holidays for 22% of the Soviet population annually. Yet by expanding nuclear power so rapidly, they foresaw an "inevitable" increase in the contamination of Ukrainian agricultural products by radionuclides, thereby endangering the health of vacationers as well as local Ukrainian residents.

The ecological damage to the republic, according to the article, was already serious and the situation was getting worse. Nuclear plants were consuming irreversibly 1.5 million cubic meters of water annually. The concentration of huge energy objects in regions with a water deficit had led to the overheating and chemical pollution of the water. Further, the regulations for the state of the water in the area of the three nuclear power plants had already been violated, wrote the scientists. Thus the water used to cool the Rovno station flows into the Styr River, and had raised its temperature by 5 degrees celsius above the admissible norm. The Horyn River, into which flows water from the Khmelnytsky station, had practically dried up downstream. The water supply in the two rivers was said to be insufficient even for providing for the capacities foreseen in the initial plans for the Rovno and Khmelnytsky stations. At South Ukraine, there was concern over the likely pollution of the water in the South Buh [Bug] River and in the Dnieper-Buh estuary. To avert the problem at Rovno and Khmelnytsky, the authorities planned "to dig up half the republic" by building an underground channel 240 kilometers in length from the Dniester River which, in the writers' view, was yet another example of the prevailing "gigantomania" and blundering tactics of the Ministry of Atomic Power.

The costs of building new reactors, it was stated, had not been worked out properly. Notably they had not taken into account the expenses involved in the protection of fuel wastes and the disassembly of the stations after their period of service. Perhaps of more importance was the siting of nuclear plants in areas subject to seismic activity and earth shifts. The authors of the article, among whom were two geologists, stated that "geological processes" that had not been provided for, such as earthquakes and seismic activity, were present in the locations of 40-70% of the area of those raions adjacent to operating and planned nuclear power plants, and that because of water diversion and other factors involved in the process of nuclear power operation, the situation was getting worse with time.

Thus, experts of the Academy of Sciences, the Ministry of Geology and State Construction of the Ukrainian SSR recently rejected the building of a transparent storage reservoir for the Rivne nuclear power plant because of the real possibility of activating volcanic-surface processes. The influence of radionuclides on the ecological parameters of the geological environment has not been studied (soils and ground water), and we do not have a prognosis of the radiological and geological influence of nuclear power plants on the environment in its entirety.

Why were the experts from the Ministry of Atomic Power ignoring such obvious factors? In the view of the Ukrainian scientists, they wanted to preserve what they perceived as their "privileged, irrefutable authority," and were thus brushing aside the warnings of scholars and the "bitter lessons" of the Chernobyl disaster. But times were changing, wrote the scientists, and it was not easy to avoid "perestroika." The theory of the "guaranteed safety" of nuclear power plant operation, with which the Soviet public had been assaulted for the past 12 years, had been proved to be erroneous. It was time for a radical change in the development of the Ukrainian economy, one that involved the transferral to resource and energy-saving technology:

> One...cannot calculate fully the moral-economic consequences of the accident at the Chernobyl nuclear power plant and be flippant about its psychological impact on the republican population. For this reason, we consider it necessary to renounce the plans for the expansion of Rivne, Khmelnytsky and South Ukraine nuclear power plants and turn to the Council of Ministers of the USSR with a petition to hear our views, and also to examine the entire complex of problems involved in the development of nuclear energy in the Ukrainian SSR.

The article, at the time of its appearance, constituted the most detailed and outspoken attack on the Ukrainian nuclear power program. It was significant also in that hitherto, the main criticisms of this program had emanated from members of the Ukrainian Writers' Union. As shown above, Soviet scientists had traditionally been scornful about the lack of knowledge of the industry among those opposing it. One suspects that it was partly because of this attitude that the Ukrainian academicians resolved to make their views known and, at the end of the article, laid out their academic credentials somewhat ostentatiously so that the Soviet authorities would not be tempted to dismiss them so easily.

There was also in the article a very definite "republican" flavor, a clear division between what was happening in discussions about nuclear power in the republic and the plans imposed from above, from the Ministry of Atomic Power of the USSR based in Moscow. The article was less emotional than that of the Cherkassy writers against the proposed Chyhyryn station described above, but nevertheless revealed great bitterness at the careless and uncaring damage to the republican ecology and environment that had resulted from poorly thought-out schemes. More subtle was the accusation that the Ministry of Atomic Power, then only 18 months old, had not been subjected to reconstruction and was trying to avoid the process. By this means, the authors attempted to equate themselves with the policies of Mikhail Gorbachev and to isolate ministry officials.

Conclusion

Have they succeeded? Lukonin has made it clear that there have been no fundamental changes to the nuclear energy plans for the future, and following the decision to abandon the Krasnodar station, he once again reaffirmed his faith in nuclear power, noting safety improvements in particular.[77] However, a serious opposition to the industry, vocal and outspoken, has clearly emerged. Today in the Soviet Union, there is a definite and growing rift between two sides on the question of nuclear energy development. Although the proponents of future expansion began with an advantage, the emergence of spokespersons such as Sakharov on the side of the opposition has swung the pendulum belatedly in the opposite direction.

Officially, nuclear energy remains a high priority in spite of and, perhaps, even because of Chernobyl. The "pro-expansion" scientists' view is that nuclear energy in the country today has been made safer, even though before April 1986, officially, no serious accidents had occurred in the industry. However, even though in this author's view, many issues of the Chernobyl disaster were deliberately concealed by the authorities, and even though an "official version" of the events appeared, it proved impossible in a period of glasnost either to conceal the real situation or to impose official thinking about nuclear energy on the Soviet public.

Eventually and with increasing confidence, the people have made their views known, and they have subsequently received powerful support. It should be stated here that the arguments for and against nuclear power are well known, but are not central to an understanding and explanation of the new developments. More important is that for the first time, an issue such as nuclear energy, which had previously not been subjected to wide discussion, was no longer a taboo topic for debate. The Ukrainians in particular, who had suffered the consequences of the accident, decided that it was time to call a halt to the development of the industry they saw as responsible for their plight. It is to the credit of Mikhail Gorbachev, although it was unlikely to have been his original intention, that an atmosphere was created that permitted the public to deal a devastating blow to the Soviet nuclear energy program.

For, despite statements of Lukonin, Legasov and others to the contrary, this is what has happened. The ambitious program cannot now be completed, if, indeed, it could have been under the best of circumstances. It lies in ruins. In Ukraine, for example, the combination of a major disaster and more open debate has seen the abandonment of one station, the cessation of work on at least two others, and acute criticism of three more, in addition to the continuing apprehension over the three remaining units of Chernobyl. This is not to say that nuclear power in the Soviet Union has no future. On the contrary, research into thermal fusion and breeder reactors is continuing all the more because of recent attacks on traditional nuclear power plants. But its development is probably a decade ahead. The current program, 1990-2000, is to all intents and purposes obsolete. Neither the RBMK nor the VVER can offer the guaranteed safety that the Soviets are seeking. Even if they could, the public has simply lost faith in the reliability and safety of the Soviet-made reactors.

This book has tried to reveal the "other side" to Chernobyl, not to play the devil's advocate, or to take the side of those in the West who have even gone so far as to accuse the USSR of an attempted "genocide" in Ukraine. The author does not doubt the bravery of those who dealt with the accident, and has praised Soviet efforts to improve their reactor design and to cooperate with the IAEA. Yet disturbing questions remain. Aspects of the disaster, such as the cleanup operation and the radioactive fallout around the plant have rarely been dealt with thoroughly or even honestly by Soviet sources. The vast majority of the proceedings of the Chernobyl Trial of July 1987 were held in secret, on the excuse—flimsy at best—that the courtroom was too small to hold the foreign and even domestic reporters. Could not a more adequate courtroom in Kiev have been chosen, given the enormous international interest in the event? It could, but the Soviet authorities clearly preferred that the trial should be held in secret.

Even when the Soviets have been obliged to abandon a reactor design because of its innate defects—the RBMK—they have been loth to admit in public that there is anything wrong with that design. Instead, the case has been made that original plans had been to develop primarily the VVER design which is common to the CMEA countries. Moreover, despite Chernobyl's impact and the long half-lives of the cesium and strontium isotopes in the radioactive fallout, not only has the land in the special zone once again been made available for agricultural use, but repopulation of the villages has occurred and is still taking place. Finally, against all logic, the Chernobyl plant's first three units have been restarted, even though a public forum in Kiev in March 1987 dashed any thoughts of expanding the station further because of the widespread public opposition to such plans and the danger of the area in question. Today, employees are working on the Chernobyl units in a zone of enhanced radiation background and before any permanent residence has been completed for the workforce. The only consolation for such people is that their days in the area may be numbered, as public protests grow.

As a postscript on the Chernobyl disaster itself, one is obliged to adhere to Oles Honchar's view, expressed so forcefully in Leningrad: science has, in this case at least, demanded its victims; for reasons of economics, psychology or international prestige. Ultimately, the value of the human element has to be measured.

NOTES

Introduction: The Cause Of The Chernobyl Accident

1. USSR State Committee on the Utilization of Atomic Energy, "The Accident at Chernobyl' Nuclear Power Plant and Its Consequences," presented at the International Atomic Energy Agency Post-Accident Review Meeting, Vienna, August 25-29, 1986.
2. USSR State Committee, "The Accident at Chernobyl"; A. Dastur, R. Osborne, D. Pendergast, D. Primeau, V. Snell and D. Torgerson, "A Quick Look at the Post-Accident Review Meeting (PARM)," *Atomic Energy of Canada Limited, AECL-9327,* September 1986.
3. P.S.W. Chan, A.R. Dastur, S.D. Grant, J.M. Hopwood, and B. Chexal, "The Chernobyl Accident: Multidimensional Simulations to Identify the Role of Design and Operational Features of the RBMK-1000," paper presented to the ENS/ANS Topical Meeting on Probabilistic Risk Assessment, Zurich, Switzerland, August 30-September 4, 1987.
4. V.G. Asmolov et al., "The Accident at the Chernobyl Nuclear Power Plant: One Year After," preliminary translation of principal Soviet paper presented by N.N. Ponomarev-Stepnoj to the International Atomic Energy Agency International Conference on Nuclear Power Plant Performance and Safety, Vienna, September 28-October 2, 1987.
5. Chan et al., "The Chernobyl Accident."
6. Ibid.

7. V.G. Asmolov et al., "The Accident at the Chernobyl Nuclear Power Plant."

8. "CANDU Nuclear Generating Station Technical Summary," *Atomic Energy of Canada Limited, PA-4,* CANDU Operations, Mississauga.

9. V.G. Snell, "Safety of CANDU Nuclear Power stations," *Atomic Energy of Canada Limited, AECL-6329,* January 1985.

10. For a detailed technical evaluation, see J.Q. Howieson and V.G. Snell, "Chernobyl—A Canadian Technical Perspective," *Nuclear Journal of Canada,* Vol. 1, no. 3 (September 1987). There is a technical executive summary of this same article in V.G. Snell and J.Q. Howieson, "Chernobyl—A Canadian Technical Perspective—Executive Summary," *Atomic Energy of Canada Limited, AECL-9334,* January 1987. For the general reader, see V.G. Snell and J.Q. Howieson, "Chernobyl—A Canadian Perspective," *Atomic Energy of Canada Limited, PA-10,* December 1986.

11. V.G. Asmolov et al., "The Accident at the Chernobyl Nuclear Power Plant."

1   The Victims Of Chernobyl

1. Interview with Uzbek journalist, Samarkand, USSR, November 1, 1987.

2. *Sovetskaya Belorussiya,* July 30, 1986.

3. *Pravda Ukrainy,* April 22, 1987.

4. *TASS,* April 28, 1987.

5. *Radyanska Ukraina,* June 13, 1987.

6. *Radio Free Europe/Radio Liberty Central News Desk,* November 20, 1986.

7. *Moscow News,* No. 46, November 16, 1986, p.12.

8. *Ibid.,* No. 29, July 19, 1987, p.4.

9. *Literaturna Ukraina,* April 29, 1987.

10. *Moscow News,* No. 32, August 9, 1987, p. 12.

11. *Associated Press,* June 6, 1987.

12. *Reuter,* July 21, 1987.

13. Interview, *Izvestiya,* Moscow, USSR, October 29, 1987.

14. *Nedelya,* No. 23, September 1986.

15. *Sovetskaya kultura,* May 13, 1986; and *Pravda Ukrainy,* May 13, 1986.

16. *Pravda Ukrainy,* May 13, 1986.

17. *News From Ukraine,* No. 16, April 1987, p.4. In his review of the author's book, *Chernobyl and Nuclear Power in the USSR* in *Slavic Review,* No. 3-4 (1987), Marshall Goldman points out correctly that the decision to transport livestock out of the irradiated zone compounded the health hazards in the area. This is another factor that can be added to the possible future casualty rate from the accident.

18. *Politicheskoe samoobrazovanie,* No. 10, October 1986, p.111.

19. *Associated Press*, September 19, 1986.
20. *Nedelya*, No. 23, September 1986, & ff.
21. *Izvestiya*, January 23, 1987.
22. *Ibid.*, January 1, 1987.
23. *Novosti*, April 14, 1987.
24. *Atomic Energy of Canada Limited (AECL)-9334*, January 1987.
25. *Novosti*, April 14, 1987, & ff.
26. *Maclean's*, July 6, 1987.
27. *Novosti*, April 14, 1987; and *Radio Free Europe/Radio Liberty Special*, August 28, 1986.
28. *Novosti*, April 14, 1987.
29. *Radio Free Europe/Radio Liberty Special*, March 31, 1987.
30. *TASS*, April 2, 1987; *Radio Kiev*, April 1, 1987; *Novosti*, April 3, 1987.
31. *Pravda Ukrainy*, April 3, 1987; *Stroitelnaya gazeta*, January 24, 1988.
32. *Pozharnoye delo*, No. 6, June 1986, pp. 24-5.
33. *Izvestiya*, September 18, 1986.
34. *Novosti*, April 23, 1987; *Moscow News*, No. 18, May 3, 1987, p.5.
35. *Reuter*, July 31, 1986.
36. *Trud*, September 11, 1986; *Robitnycha hazeta*, June 3, 1987. Various Western visitors to Kiev have confirmed that Romanenko is a far from popular figure among Kievans.
37. *Pravda Ukrainy*, October 5, 1986.
38. *Radyanska Ukraina*, October 10, 1986, & ff.
39. *Pravda*, December 15, 1986.
40. *Novosti*, April 10, 1987.
41. *TASS*, July 29, 1987.
42. *Ibid.*, February 6, 1987.
43. *Novosti*, April 10, 1987.
44. *Washington Post*, May 23, 1987 (Peter Hoffer).
45. *Maclean's*, July 6, 1987, p.7.
46. *Robitnycha hazeta*, June 3, 1987.
47. *Radio Moscow*, March 22, 1987.
48. *Sovetskaya kultura*, April 25, 1987.
49. See, for example, *Visti z Ukrainy*, No. 18, May 1987, p.3.
50. David R. Marples, *Chernobyl and Nuclear Power in the USSR* (New York: St. Martin's Press, 1986;.Edmonton: CIUS, 1986; London: Macmillan, 1987), pp. 149-50.
51. Radio Free Europe/Radio Liberty Area Audience and Opinion Research, "Soviet Background Notes: Unevaluated Comments from Recent Emigrants," *SBN* 2-87, April 1987, p.7.
52. *Newsday*, January 21, 1987 (Robert Cooke).
53. U.S. Nuclear Regulatory Commission, *Report on the Accident at the Chernobyl Nuclear Power Station, NUREG-1250*, Washington, D.C., January 1987, 8-9. (Referred to hereafter as *NUREG-1250*.)
54. *Radio Moscow*, March 22, 1987.

55. *Novosti*, April 10, 1987.
56. *TASS*, April 17, 1987, & ff.
57. *Radio Moscow*, April 20, 1987.
58. *Robitnycha hazeta*, June 3, 1987.
59. *TASS*, April 17, 1987.
60. *Sovetskaya Belorussiya*, April 26, 1987.
61. *TASS*, July 2, 1987.
62. *Ibid.*, August 19, 1987.
63. *Washington Post*, December 14, 1986.
64. *News From Ukraine*, No. 18, April 1987, p.2.
65. *Radio Kiev*, April 14, 1987.
66. *Izvestiya*, April 24, 1987.
67. *Moscow News*, No. 28, July 12, 1987, p.11.
68. *Rabotnichestvo delo*, May 19, 1987.
69. *Pravda Ukrainy*, July 16, 1987, & ff.
70. *Ibid.*
71. *Sovetskaya Belorussiya*, May 16, 1987.
72. *Associated Press*, September 30, 1986.
73. *Ibid.*, November 23, 1986.
74. *Ibid.*, June 26, 1987.
75. *Maclean's*, July 6, 1987.
76. *NUREG-1250*, 8-8 and 8-10.
77. *Radio Free Europe/Radio Liberty Special*, May 18, 1987.
78. *Ibid.*, March 27, 1987.
79. *Reuter*, May 6, 1987.
80. *TASS*, May 23, 1987.
81. *Radyanska Ukraina*, June 13, 1987.
82. *Novosti*, April 22, 1987.
83. *Izvestiya*, April 24, 1987.
84. L.A. Ilyin, O.A. Pavlovsky, "Radiologicheskie posledstviya Chernobylskoi avarii v SSSR i mery, predpriyatye s tselyu ikh smyagcheniya," *IAEA-CN-48*, Vienna, 28 September-2 October 1987; *Radio Free Europe/Radio Liberty Special*, September 30, 1987.
85. *Radio Free Europe/Radio Liberty Central News Desk*, October 15, 1987.
86. *NUREG-1250*, 8-11.

2   The Environmental Impact

1. USSR State Committee on the Utilization of Atomic Energy, *The Accident at the Chernobyl' Nuclear Power Plant and Its Consequences*, Information compiled for the IAEA Experts' Meeting, 25-29 August 1986, Vienna (August 1986), pp. 36-7. (Referred to hereafter as *The Accident at the Chernobyl' Plant.*)
2. *Radyanska Ukraina*, June 19, 1987.

3. *Radio Kiev*, July 22, 1987.
4. *Chicago Tribune*, June 15, 1987.
5. *The Accident at the Chernobyl' Plant*, p.37.
6. *Radyanska Ukraina*, June 19, 1987.
7. *Izvestiya*, April 24, 1987.
8. *Pravda Ukrainy*, July 16, 1987.
9. *Robitnycha hazeta*, June 3, 1987.
10. *The Accident at the Chernobyl' Plant*, p.37.
11. *Ibid.*
12. *DPA*, September 3, 1986.
13. *Radio Free Europe/Radio Liberty Special*, January 20, 1987.
14. *Radyanska Ukraina*, June 11, 1987.
15. *Robitnycha hazeta*, April 14, 1987.
16. *TASS*, July 8, 1986.
17. *Radyanska Ukraina*, June 19, 1987.
18. *Pravda*, October 31, 1986 & ff.
19. *Sovetskaya Belorussiya*, May 16, 1987.
20. *Moscow News*, No. 18, May 3, 1987, p..5.
21. *Pravda Ukrainy*, December 17, 1986.
22. *Ibid.*, April 5, 1987 & ff.
23. *Radio Free Europe/Radio Liberty Special*, January 20, 1987; *Radio Kiev*, —— June 16, 1987.
24. *Pravda Ukrainy*, October 10, 1986.
25. *The Globe and Mail*, June 22, 1987; *Sovetskaya Belorussiya*, May 16,_ 1987; *Robitnycha hazeta*, May 22, 1987.
26. *Reuter*, January 20, 1987.
27. *Radio Free Europe/Radio Liberty Special*, January 20, 1987.
28. *Robitnycha hazeta*, March 3, 1987, & ff.
29. *Radio Moscow*, March 10, 1987.
30. *Pravda*, March 15, 1987.
31. *Literaturnaya gazeta*, April 15, 1987.
32. *Komsomolskaya pravda*, May 1, 1987.
33. *Radio Kiev*, April 26, 1987.
34. *Radio Moscow*, April 25, 1987.
35. *Ibid.*
36. *News From Ukraine*, No. 20, May 1987, p.2
37. *Robitnycha hazeta*, June 3, 1987.
38. *TASS*, April 28, 1987.
39. *Sovetskaya Belorussiya*, July 30, 1986.
40. *Nedelya*, No. 23, September 1986.
41. *Robitnycha hazeta*, April 29, 1987.
42. *Sovetskaya Belorussiya*, April 18, 1987.
43. *TASS*, April 28, 1987; and *Pravda,* April 23, 1987.
44. *Izvestiya*, January 7, 1987.
45. *Sovetskaya Belorussiya*, May 16, 1987.

46. *Radio Free Europe/Radio Liberty Special*, January 20, 1987.

47. *Chelovek i zakon*, #10, October 1986.

48. *Pravda*, April 23, 1987.

49. *Radio Kiev*, April 14, 1987.

50. Cited by William Tuohy in *Los Angeles Times*, June 1, 1987.

51. *Sovetskaya Belorussiya*, April 18, 1987.

52. *Reuter*, August 7, 1986; and *Daily Telegraph*, July 31, 1986.

53. *NUREG-1250*, 8-16.

54. *Reuter*, August 7, 1986; and January 20, 1987.

55. *Associated Press*, June 12, 1987.

56. *Reuter*, May 12, 1987.

57. *Radio Free Europe/Radio Liberty Special*, February 4, 1987.

58. *Financial Times*, August 13, 1987.

59. *Associated Press*, October 9, 1986.

60. Cited in *Radio Free Europe/Radio Liberty Special*, March 27, 1987.

61. *Robitnycha hazeta*, March 17, 1987.

62. *Politicheskoe samoobrazovanie*, #10, October 1986.

63. *Sovetskaya Rossiya*, July 18, 1986, & ff.

64. *Robitnycha hazeta*, March 17, 1987. A shortened version of this interview was published in the Kiev newspaper for Ukrainians abroad, *Visti z Ukrainy*, No. 15, 1987.

65. *Sovetskaya Belorussiya*, April 18, 1987.

66. *Pravda Ukrainy*, April 3, 1987, & ff. The report was provided by V. Krut, a doctor of agricultural studies, and B. Prister, a doctor of biological studies.

67. *Sovetskaya Belorussiya*, April 18, 1987, & ff. Although it was considered somewhat dangerous to grow buckwheat, it has been pointed out elsewhere that buckwheat "removes radioactive particles from the organ." In April 1987, the Ukrainian authorities were complaining that it was almost impossible to find it other than at the collective farm market, where it was very expensive, and that harvests of the crop had been very poor. In short, therefore, it must have been recommended as a source to combat the effects of radiation on the human organism. See *Literaturna Ukraina*, April 9, 1987.

68. *Radyanska Ukraina*, June 13, 1987.

69. *Moscow News*, No. 28, July 19, 1987, p.11.

70. *Sovetskaya Belorussiya*, May 16, 1987.

71. Earlier measures against dust are mentioned in the author's 1986 study, *Chernobyl and Nuclear Power in the USSR*, p. 160.

72. *Izvestiya*, April 24, 1987; and *Radio Kiev*, June 16, 1987.

73. *Sovetskaya Belorussiya*, April 18, 1987, & ff.

74. *Sotsialisticheskaya industriya*, April 26, 1987.

75. *Novosti*, April 21, 1987.

76. The interview with Anatoly Duda, a Deputy Chairman of the Propaganda Department in the Kiev Oblast Executive Committee was cited by Mary Ellen Bortin of *Reuter agency*, June 17, 1987.

77. *Sovetskaya Belorussiya*, July 15, 1987.
78. *TASS*, July 9, 1987.

3 The Economic And Political Repercussions

1. *Sotsialisticheskaya industriya*, September 30, 1986.
2. *Izvestiya*, October 3, 1986.
3. *Radio Moscow*, October 5, 1986.
4. Interview, Moscow, October 27, 1987.
5. *Sovetskaya Rossiya*, November 27, 1986.
6. *Ekonomicheskaya gazeta*, No. 45, November 1986, p.3.
7. *TASS*, October 11, 1986.
8. *Sovetskaya Belorussiya*, January 6, 1987. It was implicit from the article that the second unit of the power station was already in service.
9. *Radio Moscow*, April 16, 1987.
10. *Izvestiya*, October 12, 1986.
11. *Ibid*.
12. *Izvestiya*, October 14, 1986.
13. *Radio Moscow*, December 11, 1986.
14. *Ekonomicheskaya gazeta*, No. 11, March 1987, p.3.
15. *Ibid.*, No. 18, May 1987, p.3.
16. *MTI*, October 31, 1986.
17. *The Wall Street Journal*, December 4, 1986.
18. *MTI*, November 5, 1986.
19. "Hungary Complains About CMEA Electric Power Supply," *Radio Free Europe, RAD/East* (Okolicsanyi), November 6, 1986.
20. *Radio Prague*, November 17, 1986; and *Radio Budapest*, November 29, 1986.
21. *Radio Free Europe/Radio Liberty Special*, January 20, 1987.
22. *Ibid.*, February 9, 1987.
23. *Radio Free Europe, RAD Background Report/67* (Girnius), April 24, 1987.
24. *TASS*, August 19, 1987.
25. *Novosti*, March 17, 1987.
26. *Sovetskaya Rossiya*, April 26, 1987.
27. *Radio Moscow*, April 25, 1987.
28. See, for example, the interview with the newly appointed Chernobyl plant Director, Erik Pozdyshev, in *Pravda*, October 10, 1986.
29. *Robitnycha hazeta*, October 21, 1987.
30. *Sovetskaya Rossiya*, April 30, 1987.
31. *Sotsialisticheskaya industriya*, April 26, 1987.
32. *Frankfurter Allgemeine Zeitung*, July 8, 1987.
33. *TASS*, November 3, 1986.
34. *Slovo lektora*, #3, March 1987, p.26.

35. *TASS*, April 22, 1987. See also, N.I. Lukonin, "Nuclear Power After Chernobyl': Current Problems and Future Indices of Nuclear Power Plants," IAEA, Vienna, 28 September-2 October 1987, *IAEA-CN-48/269*.

36. *Radio Moscow*, July 27, 1987.

37. See Marples, *Chernobyl and Nuclear Power in the USSR*, pp. 103-6.

38. *Sotsialisticheskaya industriya*, October 31, 1986.

39. *Radio Moscow*, July 21, 1987, & ff.

40. *Robitnycha hazeta*, October 21, 1987.

41. *Radyanska Ukraina*, December 3 and 5, 1986.

42. The accident at Rovno, which has been mentioned by the former Ukrainian dissident Iosyp Terelya, was hinted at by Andrei Pralnikov in the most outspoken Soviet newspaper, *Moscow News*, No. 23, August 9, 1987, p.12. No other sources have made reference to such an accident.

43. "Poland Announces A Further Delay On Its Nuclear Energy Program," *Radio Free Europe*, *RAD-East*/Stefanowski, November 11, 1986.

44. Could this concern have been sparked by the events at Rovno? It is odd that two references to structural problems should be made at plants located relatively close to each other at almost the same time.

45. *Business Magazine*, February 2, 1987.

46. *PAP*, April 9, 1987.

47. *Radio Warsaw*, August 17, 1987. [1105GMT]

48. *Ibid.* [0730GMT]

49. See Marples, *Chernobyl and Nuclear Power in the USSR*, p.89.

50. This is a familiar criticism. Similar reports have emanated from the South Ukraine nuclear plant, for example, an interview with brigadier A.I. Chernikova, *Radio Kiev*, July 9, 1987.

51. *Radyanska Ukraina*, May 15, 1987, & ff.

52. *Molod Ukrainy*, July 4, 1987, & ff.

53. See, for example, *Molod Ukrainy*, July 1, 1987 on the South Ukraine nuclear power plant.

54. *Robitnycha hazeta*, October 21, 1987. The interviewee was F.S. Temirov, Director of the all-Union planning institute "Atomenergostroiproekt" [Atomic energy construction planning].

55. Interview with editorial board of *Izvestiya*, Moscow, November 27, 1987.

56. See Marples, *Chernobyl and Nuclear Power in the USSR*, Chapter 6.

57. *Pravda Ukrainy*, August 7, 1987.

58. See Roman Solchanyk, "Chernobyl: The Political Fallout in Ukraine," *Journal of Ukrainian Studies* 20 (Summer 1986): 20-34, and especially 28-32.

59. At the time of writing, the most recent dismissal was that of the Volyn Oblast First Party Secretary, Zinovii Kovalchuk, *Radio Moscow*, July 4, 1987.

60. *Komsomolskaya pravda*, December 7, 1986.

61. *Pravda*, October 10, 1986.

62. *Reuter*, June 18, 1987.
63. *TASS*, July 20, 1986.
64. *Robitnycha hazeta*, October 21, 1987.
65. *Pravda*, December 16, 1986.
66. *Radio Free Europe/Radio Liberty Special*, July 8, 1987.
67. Interview, *Izvestiya*, Moscow, November 27, 1987.
68. Keith Bush, "Glasnost and the Chernobyl Trial," *Radio Liberty Research*, F-546, July 29, 1987.
69. *Moscow News*, No. 29, July 19, 1987, p.4.
70. *Izvestiya*, July 8, 1987.
71. *Robitnycha hazeta*, August 1, 1987.
72. *Moscow News*, No. 32, August 9, 1987, p.12.
73. *News From Ukraine* No. 32, August 1987, p.5.
74. *Moscow News*, No. 32, August 9, 1987, p.12.
75. *Robitnycha hazeta*, August 1, 1987.
76. *Moscow News*, No. 29, July 19, 1987, p.4.
77. *News From Ukraine*, No. 32, August 1987, p.5.
78. *Associated Press*, July 7, 1987.
79. *The Daily Telegraph*, July 8, 1987.
80. *Le Matin* (Lausanne, Switzerland), July 8, 1987.
81. *Robitnycha hazeta*, August 1, 1987.

4   Images Of Chernobyl: Arts And The Public

1. *Radio Kiev*, July 21, 1987.
2. *Komsomolskoye znamya*, March 13, 1987.
3. *Pravda Ukrainy*, June 12, 1987.
4. *Ibid.*, June 11, 1987, & ff.
5. *Sovetskaya Litva*, July 26, 1987.
6. *Pravda Ukrainy*, June 12, 1987.
7. *Ibid.*
8. *Radio Moscow*, April 5, 1987.
9. Interview with Dr. Harold Denton, Office of Nuclear Reactor Safety, U.S. Nuclear Regulatory Commission, Washington, D.C., April 23, 1987.
10. *Moscow News*, No. 17, April 26, 1987, p.13.
11. *Reuter*, April 24, 1987.
12. *Pravda Ukrainy*, June 12, 1987.
13. *Reuter*, April 13, 1987.
14. *The Sunday Times*, April 19, 1987.
15. *Literaturnaya gazeta*, June 3, 1987.
16. *Ibid.* (E. Alekseev)
17. *Sovetskaya kultura*, April 25, 1987, pp.4-5.
18. *The Independent*, March 4, 1987.

19. *Literaturnaya gazeta*, June 3, 1987.
20. *Sovetskaya kultura*, April 25, 1987, pp.4-5.
21. *Radio Moscow*, March 21, 1987.
22. *Pravda Ukrainy*, April 3, 1987.
23. The following comments are based on the interview that appeared in *Robitnycha hazeta*, April 11, 1987.
24. See, for example, *Robitnycha hazeta*, April 3, 1987. Initially there did not seem to be any doubts regarding the cause of Shevchenko's death. Only later did Soviet scientists deny that radiation sickness had been the cause. Gradually the later "diagnoses" superseded the earlier ones in all Soviet accounts and today, the idea that the Ukrainian film director died of radiation poisoning is attributed to "malicious rumors," despite the fact that these rumors emanated from Soviet accounts in the first place. In the view of this author, there is no reason to doubt the first reports. The later denials might stem from the desire to keep the official death toll stationary at 31. By the spring of 1987, radiophobia was the main concern of the authorities. Additional victims would only have exacerbated this very serious trend.
25. *Robitnycha hazeta*, April 11, 1987.
26. *Pravda Ukrainy*, April 3, 1987.
27. *Kultura i zhyttya*, April 12, 1987; and *Literaturnaya gazeta*, June 3, 1987.
28. *Pravda*, July 3, 1987.
29. *Politicheskoe samoobrazovanie*, #10, October 1986, p.112.
30. *Izvestiya*, March 6, 1987.
31. *Komsomolskoye znamya*, March 13, 1987.
32. *Radyanska Ukraina*, March 28, 1987.
33. *Komsomolskoye znamya*, March 11, 1987.
34. *Ibid.*, March 13, 1987.
35. *Noorte haal*, August 13, 1986. The author does not read Estonian and hence relied on the translation of these articles that appeared in *FBIS*.
36. *Literaturna Ukraina*, May 21, 1987.
37. *The Washington Post*, December 14, 1986.
38. *Komsomolskoye znamya*, March 11, 1987.
39. *Izvestiya*, January 12, 1987.
40. *Pravda*, January 20, 1987.
41. *Molod Ukrainy*, January 1, 1987.
42. *News From Ukraine*, No. 16, April 1987, p.1.
43. *Izvestiya*, January 7, 1987; and *Komsomolskoye znamya*, March 13, 1987.
44. *Molod Ukrainy*, January 1, 1987.
45. *Komsomolskoye znamya*, March 11, 1987. It is interesting that Chernihiv was also considered a contaminated zone. It had not been evacuated and there were no official reports to suggest that it may have been affected seriously by radioactive fallout.
46. *Politicheskoe samoobrazovanie*, #10, October 1986, p.112.

47. *Ibid.*, pp. 113-14.
48. *Literaturna Ukraina*, May 21, 1987.
49. *Radyanska Ukraina*, April 26, 1987.
50. *Komsomolskaya pravda*, May 1, 1987.
51. *Associated Press*, June 18, 1987.
52. *Literaturna Ukraina*, October 7, 1987.
53. *Pravda*, August 8, 1986; *Pravda Ukrainy*, August 7, 1986.
54. *Chelovek i zakon*, #10, October 1986.
55. *Radio Moscow*, November 5, 1986.
56. *Radyanska Ukraina*, January 1, 1987.
57. *Pravda Ukrainy*, March 18, 1987; *Izvestiya*, March 9, 1987. The *Daily Star* is not to be confused with the British Communist Party newspaper, the *Morning Star*. This error was made by Felicity Barringer writing in the *New York Times* of April 6, 1987. In fact, it is a conservative newspaper that tends to support the policies of the Thatcher government.
58. *Pravda Ukrainy*, March 18, 1987.
59. *TASS*, April 25, 1987.
60. *Radio Moscow*, May 8, 1987.
61. This statement also applies to groups that are very anti-Soviet in makeup. In Canada, for example, the Ukrainian Self-Reliance League drafted a dedication for a plaque that was to commemorate the first victims of the disaster (all firemen), which at first was to have read (before it was amended) "To the memory of those firemen who died in Russian-occupied Ukraine...."
62. *Pravda Ukrainy*, June 12, 1987.
63. *Associated Press*, June 30, 1987; *Radio Free Europe/Radio Liberty Central News Desk*, July 4, 1987.
64. *Pravda Ukrainy*, August 1, 1987.
65. *Ibid.*, April 22, 1987.
66. *Voyennyi vestnik*, October 24, 1986.
67. *Komsomolskoye znamya*, October 16, 1986, & ff.
68. *Molod Ukrainy*, January 1, 1987, & ff.
69. *Pravda*, October 27, 1987.
70. *Pravda Ukrainy*, September 16, 1986.
71. *Chelovek i zakon*, #10, October 1986.
72. *Sovetskaya Rossiya*, October 13, 1986.
73. *Robitnycha hazeta*, January 9, 1987.
74. *Ibid.*, December 27, 1986.
75. *Visti z Ukrainy*, No. 19, May 1987, p.7.
76. *Agitator*, #13, July 1987.

5   The Special Zone

1. *Novosti*, April 10, 1987.
2. *Robitnycha hazeta*, March 17, 1987.
3. *Izvestiya*, April 24, 1987.
4. *New Times*, No. 29, July 28, 1986, p.15.
5. *Soviet Television/B*, September 15, 1986.
6. *Pravda*, November 15, 1986.
7. *Ibid.*, December 16, 1986.
8. *Komsomolskaya pravda*, October 3, 1986.
9. *Izvestiya*, December 30, 1986.
10. *Krymska pravda*, October 2, 1986, & ff.
11. *Izvestiya*, December 30, 1986.
12. *Komsomolskaya pravda*, December 7, 1986.
13. *Robitnycha hazeta*, December 16, 1986; and March 1, 1987.
14. *Ibid.*, December 16, 1986.
15. *Pravda Ukrainy*, November 1, 1986.
16. *Trud*, January 13, 1987.
17. *Pravda*, March 6, 1987.
18. *Radyanska Ukraina*, June 11, 1987, & ff.
19. *Komsomolskaya pravda*, December 7, 1986. See also Toomas Ilves, "Estonians Help At Chernobyl," *Radio Free Europe, RAD Background Report*, September 10, 1986.
20. *New Times*, No. 29, July 28, 1986, p.15.
21. *CETEKA*, May 18, 1987, citing the *Novosti* Press Agency.
22. *Izvestiya*, January 7, 1987.
23. *Radio Free Europe/Radio Liberty Special*, January 21, 1987.
24. *Robitnycha hazeta*, March 13, 1987, & ff.
25. *Chelovek i zakon*, #10, October 1986.
26. *Radyanska Ukraina*, October 9, 1986, & ff.
27. *Ibid.*
28. *Robitnycha hazeta*, January 23, 1987.
29. All the following entries are taken from *Komsomolskaya pravda*, October 24, 1986.
30. This signifies that the students were of the 17-18 years age group. Presumably the entry in question was written by a 16-year-old or younger.
31. One can assume that the young woman was not living permanently in her Prypyat apartment. It is probable that she had returned in August 1986 to check on her goods, which explains why the letter to her husband would still have been in place three months after it was written.
32. It is quite common for Ukrainians to work on the construction of East European nuclear reactors, most notably the Paks station in Hungary and Kozloduy in Bulgaria. There are also workers from the USSR at the Zarnowiec nuclear plant being built in the Gdansk area of Poland.
33. Prypyat.

34. This is the White Steamer (Bilyi Paroplav in Ukrainian) moored on the Kiev Reservoir that was used for several months as temporary accommodation for those building Zelenyi Mys and for the building workers who at that time were to have continued their work on the fifth RBMK-1000 at Chernobyl. Their work was postponed indefinitely in April 1987.

35. *Lesnaya promyshlennost*, August 30, 1986.

36. *Radio Kiev*, October 16, 1986.

37. *Reuter*, June 18, 1987. Interview with Oleksandr Kovalenko, the spokesperson for the Kombinat production association.

38. *Robitnycha hazeta*, April 10, 1987.

39. *Radio Moscow*, April 24, 1987.

40. *Reuter*, June 18, 1987.

41. Dzintra Bungs, "Latvian Aid Helps Deal With Effects of the Chernobyl Accident," *Radio Free Europe/Baltic Area SR/4*, July 18, 1986.

42. *Leninskoye znamya*, August 20, 1986.

43. *Robitnycha hazeta*, April 10, 1987.

44. *Ibid.*, November 18, 1986.

45. *News From Ukraine*, No. 3, January 1987, p.2.

46. *Robitnycha hazeta*, October 24, 1986.

47. *Krasnaya zvezda*, October 11, 1986.

48. *Robitnycha hazeta*, October 24, 1986, & ff.

49. *Radio Kiev*, October 16, 1986.

50. *Radio Free Europe Central News Desk/Latvian BD*, November 14, 1986.

51. Cited by Dzintra Bungs, "Soviet Latvian Press on Chernobyl: More Optimistic (But Flawed) Reports," *Radio Free Europe RAD/East*, November 14, 1986.

52. *News From Ukraine*, No. 16, April 1987, p.4.

53. *Pravda*, December 15, 1986.

54. *Krasnaya zvezda*, March 13, 1987.

55. *Reuter*, June 18, 1987.

56. *Krasnaya zvezda*, March 13, 1987.

57. *Radio Moscow*, March 22, 1987.

58. *Novosti*, April 14, 1987.

59. *Noorte haal*, August 14, 1986, & ff.

60. *Ibid.*, August 16, 1986, & ff.

61. *Ibid.*, August 14, 1986. Ilves cites tourists' accounts that the men were initially told that their term in the special zone would be limited to 30 days.

62. Report of the Relief Center for Estonian Prisoners of Conscience in the USSR, Stockholm, October 30, 1986; *Neuer Zuercher Zeitung*, November 2-3, 1986.

63. Cited by *Moscow News*, No. 20, May 17, 1987, p.2.

64. One assumes that the author is referring to the Moscow Trust Group, an environmental group that has also protested against nuclear power in Moscow.

65. *Sovetskaya kultura*, August 8, 1987.
66. *Ibid.*
67. *Moscow News*, No. 20, May 17, 1987, p.2., & ff.
68. *Molodezh Estonii*, April 30, 1987.
69. *Tartu Edasi*, July 15, 1986, as cited by Toomas Ilves, "Additional Information on Estonians at Chernobyl," *Radio Free Europe, RAD/East*, October 14, 1986.
70. Toomas Ilves, "CC Propaganda Chief Reportedly Dismissed for Chernobyl Articles," *Radio Free Europe, Baltic Area SR/4*, June 15, 1987.
71. *Radio Stockholm*, July 6, 1987. [1700 GMT]
72. *Reuter*, July 6, 1987.
73. *Radio Stockholm*, July 6, 1987.
74. *Radio Moscow*, July 15, 1987.
75. *Sovetskaya Estoniya*, November 14, 1986.

6   Restoration And Reconstruction

1. *Noorte haal*, August 13, 1986.
2. *Pravda*, September 1, 1986.
3. *Politicheskoe samoobrazovanie*, #10, October 1986, p.116. The statement is curious in that the two villages in question are on the outskirts of the 30-kilometer zone. It was noted earlier that it was considered feasible for two elderly women to go back to Ladyzhychi, which is closer to the damaged reactor than the villages cited. It could indicate the "spottiness" of the radioactive fallout, however.
4. *Ibid.*, pp. 116-17, & ff.
5. *Selskaya zhizn*, August 17, 1986.
6. *Izvestiya*, August 10, 1986.
7. *Pravda Ukrainy*, October 31, 1986, & ff.
8. *Pravda*, December 16, 1986.
9. *Izvestiya*, January 9, 1987.
10. *Radio Moscow*, April 12, 1987; *Pravda Ukrainy*, April 22, 1987.
11. *Pravda Ukrainy*, August 7, 1987.
12. *Radyanska Ukraina*, April 26, 1987.
13. *Pravda Ukrainy*, June 12, 1987. Incidentally, the 250 homes constructed could hardly have accounted for all the former residents of Zalissya, since the village had over 3,000 residents in the late 1960s. See *Istoriya mist i sil Ukrainskoi RSR: Kyivska Oblast* (Kiev, 1971), p.724.
14. *The Times*, June 11, 1987.
15. *Associated Press*, June 18, 1987.
16. *Radyanska Ukraina*, March 28, 1987, & ff.
17. *Ibid.*
18. *Radio Moscow*, August 29, 1986.
19. *Izvestiya*, September 23, 1986.

20. *TASS*, September 29, 1986; *Radio Kiev*, September 30, 1986; *Radio Moscow*, October 1, 1986.
21. *Izvestiya*, October 3, 1986.
22. *Pravda Ukrainy*, October 3, 1986.
23. *Radio Moscow*, October 2, 1986.
24. *Ibid.*, October 1, 1986.
25. *Pravda*, October 10, 1986.
26. *Robitnycha hazeta*, December 16, 1986.
27. *Izvestiya*, November 10, 1986.
28. *Robitnycha hazeta*, April 29, 1987.
29. *News From Ukraine*, No. 19, May 1987, p.2.
30. *Pravda*, October 10, 1986.
31. *Pravda Ukrainy*, November 8, 1986.
32. *Radio Moscow*, April 25, 1987.
33. *Sovetskaya Belorussiya*, May 16, 1987.
34. *United Press International*, August 17, 1987.
35. Interview, *Izvestiya*, Moscow, November 27, 1987.
36. *Izvestiya*, December 5, 1987; and *Radyanska Ukraina*, December 6, 1987.
37. *Visti z Ukrainy*, No. 15, April 1987.
38. *News From Ukraine*, No. 16, April 1987, p.4.
39. *Literaturnaya gazeta*, May 21, 1987, & ff.
40. *Moscow News*, No. 17, April 26, 1987, p.13, & ff.
41. *Literaturnaya gazeta*, May 21, 1987, & ff.
42. *Pravda Ukrainy*, April 14, 1987.
43. *Sovetskaya kultura*, April 25, 1987.
44. *Moscow News*, No. 17, April 26, 1987, p.13.
45. *Robitnycha hazeta*, December 16, 1986.
46. *News From Ukraine*, No. 16, April 1987, p.4.
47. *Moscow News*, No. 17, April 26, 1987, p.13.
48. *Radio Kiev*, April 28, 1987.
49. *Literaturna Ukraina*, May 21, 1987.
50. *Kyivska pravda*, June 28, 1987.
51. *Ibid.*
52. *Sotsialisticheskaya industriya*, November 6, 1986; *Robitnycha hazeta*, November 18, 1986.
53. *Robitnycha hazeta*, December 16, 1986.
54. *Izvestiya*, April 24, 1987; *Sovetskaya Belorussiya*, May 16, 1987.
55. *Radio Moscow*, May 14, 1987.
56. *Radyanska Ukraina*, June 11, 1987.
57. *Pravda Ukrainy*, June 12, 1987.
58. *Ibid.*
59. *Radio Moscow*, May 14, 1987.
60. Some of the most vivid descriptions of Prypyat were provided by William J. Eaton. See the *Los Angeles Times*, June 16, 1987 and July 6, 1987.

Soviet journalists resented the fact that Eaton had taken photographs of washing hanging from the balconies of the deserted Prypyat apartments. See, e.g., *Pravda Ukrainy*, June 12, 1987.

61. *Moscow News*, No. 28, July 12, 1987, p.11.
62. *Pravda*, December 15, 1986, & ff.
63. *Izvestiya*, January 17, 1987, & ff.
64. *Pravda Ukrainy*, April 22, 1987.
65. *Reuter*, April 24, 1987.
66. *Literaturna Ukraina*, May 21, 1987.
67. *Associated Press*, June 18, 1987.
68. *Pravda*, July 23, 1986.
69. *Ibid.*, August 8, 1986; *Sovetskii patriot*, August 24, 1986.
70. See David R. Marples, "Chernobyl' and Ukraine," *Problems of Communism*, No. 6 (November-December 1986): 24.
71. *Pravda*, October 20, 1986.
72. *Izvestiya*, October 3, 1986
73. *Trud*, October 21, 1986.
74. *Radyanska Ukraina*, October 3, 1986.
75. *Trud*, October 21, 1986, & ff.
76. *Robitnycha hazeta*, December 25, 1986.
77. *Ibid.*
78. *Pravda*, October 10, 1986.
79. *Robitnycha hazeta*, November 18, 1986; *Pravda Ukrainy*, December 16, 1986.
80. *Agitator*, #13, July 1987.
81. *The Times*, June 11, 1987.
82. *Robitnycha hazeta*, October 26, 1986.
83. *Pravda Ukrainy*, January 1, 1987.
84. *Robitnycha hazeta*, January 10, 1987.
85. *Stroitelnaya gazeta*, September 2, 1987. *Pravda Ukrainy*, January 1, 1987.
86. *Pravda Ukrainy*, January 1, 1987.
87. *Radyanska Ukraina*, January 1, 1987.
88. *Pravda Ukrainy*, January 1, 1987.
89. *Ibid.*, December 14, 1986; and December 16, 1986.
90. *Radyanska Ukraina*, January 1, 1987; *Robitnycha hazeta*, October 26, 1986, and April 25, 1987.
91. *Robitnycha hazeta*, January 10, 1987.
92. *Izvestiya*, March 10, 1987.
93. *Radio Kiev*, April 23, 1987.
94. *Robitnycha hazeta*, April 25, 1987, & ff.
95. *Ibid.*, June 26, 1987.
96. *Radio Kiev*, April 29, 1987.
97. The following comments are based on the account in *Robitnycha hazeta*, April 25, 1987.

98. *Molod Ukrainy*, June 20, 1987, & ff.
99. *Robitnycha hazeta*, July 25, 1987, & ff.
100. *Ibid.*
101. *Ibid.*
102. *Pravda Ukrainy*, March 18, 1987; *Radyanska Ukraina*, April 11, 1987.
103. *Robitnycha hazeta*, June 26, 1987.
104. *Ibid.*, July 25, 1987.
105. *Molod Ukrainy*, June 20, 1987.
106. *Robitnycha hazeta*, June 26, 1987.
107. *Moscow News*, No. 32, August 9, 1987, p.12.
108. *Pravda Ukrainy*, September 17, 1987.
109. *Radyanska Ukraina*, November 24, 1987. *Kultura i zhyttya*, December 6, 1987, cites a concert being held for shiftworkers at Zelenyi Mys, who were to be moved to the new city of Slavutych "in a short while."
110. *Pravda Ukrainy*, November 19, 1987, & ff.
111. *Radyanska Ukraina*, November 24, 1987.
112. *Sotsialisticheskaya industriya*, December 4, 1987. Incidentally, in the author's view, the Western news agencies did not misinterpret this account. The way in which it was written up in the newspaper, by Special Correspondent, L. Sotnik, implied that the "incidents" had occurred at the nuclear plant itself at the fuel assembly unit of Chernobyl-2, at which "gross violations" had taken place. Not one example was given of such a violation at Slavutych in this article (although we know, of course, that many had occurred over the course of 1987). Consequently, and given the reprimand to Umanets, the conclusions of the Western reporters *were accurate given the information provided by the Soviet account*. In conclusion, at the very least, it is evident that the deaths had been concealed for some time, unless they were mentioned in local newspapers at the oblast and raion level and were thus unavailable in the West.
113. *Moscow News*, No. 51, December 20, 1987, p.6.
114. This was noted in the speech by the Ukrainian writer Oles Honchar in Leningrad, as reported in *Literaturna Ukraina*, October 7, 1987.

7   The Nuclear Power Debate

1. *Tanjug*, May 21, 1987.
2. *Radio Kiev*, April 2, 1987.
3. Unless otherwise stated, the following comments are based on *Visnyk Akademii nauk Ukrainskoi RSR*, #10, October 1983. [Hereafter abbreviated to *Visnyk AN URSR*.]
4. One gigawatt is equal to 1 billion watts.
5. *News From Ukraine*, No. 32, August 1987, p.3.
6. *Visnyk AN URSR*, #10, October 1983.

7. *News From Ukraine*, No. 32, August 1987, p.3.
8. *Radio Kiev*, October 26, 1986.
9. *Radio Moscow*, July 29, 1987, & ff.
10. *Robitnycha hazeta*, July 31, 1987.
11. *Ibid.*, August 21, 1987.
12. *Izvestiya*, January 14 and 15, 1987.
13. *News From Ukraine*, No. 4, January 1987, p.2.
14. "Report from the Commission to the Council and the European Parliament. The Chernobyl Nuclear Power Plant Accident and Its Consequences in the Framework of the European Community," *Commission of the European Communities*, October 1986, p.23.
15. International Atomic Energy Agency, "Convention on Early Notification of a Nuclear Accident, 26 September 1986," cited in *Survival* (May-June 1987): 268-71.
16. *Ibid.*, pp. 271-72.
17. *TASS*, November 14, 1986.
18. *Survival* (May-June 1987): 270.
19. *Pravda Ukrainy*, May 14, 1987.
20. *Sovetskaya Rossiya*, April 26, 1987.
21. *Pravda Ukrainy*, May 14, 1987.
22. *The Chernobyl Accident: A Year Later* (Soviet report to the IAEA), September-October 1987, Section 5-2.
23. *Commission of the European Communities*, October 1986, p.18.
24. *The Chernobyl Accident: A Year Later*, Section 5-2.
25. *Trud*, January 13, 1987.
26. *Radio Moscow*, April 25, 1987.
27. *Pravda Ukrainy*, May 14, 1987.
28. *Izvestiya*, June 22, 1987.
29. *Radio Kiev*, June 15, 1987.
30. *Sotsialisticheskaya industriya*, July 31, 1987.
31. *Ibid.*, April 26, 1987.
32. *Moscow News*, No. 32, August 9, 1987, p.12.
33. *Pravda Ukrainy*, December 16, 1986.
34. *News From Ukraine*, No. 16, April 1987, p.4.
35. *Pravda*, April 5, 1987.
36. *Radio Kiev*, April 20, 1987.
37. *TASS*, May 11, 1987.
38. N.F. Lukonin, "Nuclear Power After Chernobyl': Current Problems and Future Indices of Nuclear Power Plants," International Atomic Energy Agency, Vienna. 28 September-2 October 1987, *IAEA-CN-48/269*, p.9.
39. Cited by *CETEKA*, May 18, 1987.
40. *Sovetskaya Rossiya*, June 6, 1987.
41. *Sovetskaya Belorussiya*, May 16, 1987.
42. *Robitnycha hazeta*, April 29, 1987.
43. *TASS*, May 11, 1987.

44. Lukonin, "Nuclear Power After Chernobyl'," pp. 8-9.
45. *Pravda*, October 27, 1986, & ff.
46. *Trud*, January 13, 1987.
47. *Novosti*, March 17, 1987.
48. *Trud*, January 13, 1987.
49. *TASS*, April 18, 1987.
50. *Trud*, January 13, 1987.
51. *Novosti*, March 17, 1987; *Slovo lektora*, #3, March 1987, p.25.
52. *Robitnycha hazeta*, January 16, 1987.
53. See, for example, David Marples, "The Kuznetsk Alternative," *Radio Liberty Research Bulletin*, RL 413/87, October 21, 1987, which notes the coal-mining tragedies in the Ukrainian coalfield at the Yasinivka-Hlyboka and Chaikino mines, in December 1986 and May 1987 respectively
54. *Slovo lektora*, #3 March 1987, p.24.
55. *Izvestiya*, April 24, 1987.
56. *Slovo lektora*, #3, March 1987, p.25.
57. *The Armenian Reporter*, November 27, 1986.
58. *Trud*, January 13, 1987.
59. *Radio Liberty/Armenian BD*, March 30, 1987.
60. *Sovetskaya Rossiya*, April 26, 1987.
61. *Ibid.*, April 30, 1987.
62. *Moscow News*, No. 28, July 12, 1987, p.11.
63. *Literaturna Ukraina*, August 6, 1987, & ff.
64. *Soviet Geography*, October 1985, pp. 643, 646.
65. *Istoriya mist i sil Ukrainskoi RSR: Cherkaska Oblast* (Kiev, 1972), p.697.
66. *Literaturna Ukraina*, August 6, 1987, & ff.
67. *Ibid.*, October 7, 1987.
68. The question was raised by the author, who was present at the press conference in Moscow, on October 28, 1987. Earlier Gurii Marchuk, President of the Academy of Sciences of the USSR, had been emphasizing the importance of preserving the country's ecology, a point that Honchar had also raised. In fairness to Velikhov, he did elaborate on the safety improvements made to Soviet reactors in the light of Chernobyl.
69. *Robitnycha hazeta*, October 21, 1987.
70. *The Christian Science Monitor*, February 4, 1988.
71. *Sovetskaya kultura*, January 5, 1988.
72. *Molod Ukrainy*, January 4, 1988.
73. Legasov's comments were carried by *DPA*, November 22, 1987. On December 22, 1987, *TASS* announced the startup of Zaporizhzhya-4, while the operation of Khmelnytsky-1 was noted by *Radyanska Ukraina*, December 23, 1987. The stoppage of work on and abandonment of the Krasnodar project was revealed in *Pravda*, January 21, 1988, and elaborated upon in *Komsomolskaya pravda*, January 27, 1988.
74. *Moscow News*, No. 2, January 10, 1988, p.10.

75. *TASS*, February 10, 1988, citing *Moscow News*, No. 7, February 14, 1988.

76. *Literaturna Ukraina*, January 21, 1988, & ff. A conciliatory but relatively ineffective response to this article was given by representatives of the nuclear power industry in *Pravda Ukrainy*, February 21, 1988.

77. *TASS*, February 9, 1988.

REFERENCES

Soviet Sources

Agitator
Chelovek i zakon
Ekonomicheskaya gazeta
Ekonomika Radyanskoi Ukrainy
Istoriya mist i sil Ukrainskoi RSR: Cherkaska Oblast (Kiev, 1972); Kyivska
    Oblast (Kiev, 1971).
Izvestiya
Komsomolskaya pravda
Komsomolskoye znamya
Krasnaya zvezda
Krymska pravda
Kultura i zhyttya
Kyivska pravda
Leninskoye znamya
Lesnaya promyshlennost
Literaturna Ukraina
Literaturnaya gazeta
Molod Ukrainy
Molodezh Estonii
Moscow News
Nedelya

New Times
News From Ukraine
Novosti
Politicheskoe samoobrazovanie
Pozharnoye delo
Pravda
Pravda Ukrainy
V.P. Pronin, *Pyatiletnyi plan energetikov respubliki* (Kiev, 1986)
Radio Kiev
Radio Moscow
Radyanska Ukraina
Robitnycha hazeta
Selskaya zhizn
Silski visti
Slovo lektora
Sotsialisticheskaya industriya
Sovetskaya Belorussiya
Sovetskaya Estoniya
Sovetskaya kultura
Sovetskaya Litva
Sovetskaya Rossiya
Sovetskii patriot
Stroitelnaya gazeta
TASS
Trud
USSR State Committee for the Utilization of Atomic Energy
Visnyk Akademii nauk Ukrainskoi RSR
Visti z Ukrainy
Voyennyi vestnik

Non-Soviet Sources

Armenian Reporter
Associated Press
Atomic Energy of Canada Limited
Business Magazine
Chicago Tribune
Commission of the European Communities
Daily Telegraph
Financial Times
Frankfurter Allgemeine Zeitung
The Globe and Mail
The Independent
International Atomic Energy Agency

Journal of Ukrainian Studies
Le Matin (Lausanne, Switzerland)
Maclean's
The New York Times
Neuer Zuercher Zeitung
Newsday
Problems of Communism
Radio Budapest
Radio Free Europe/Radio Liberty
Radio Liberty Research Bulletin
Radio Prague
Radio Stockholm
Radio Warsaw
Reuter
Soviet Analyst
Soviet Geography
The Sunday Times
Survival
Tanjug
Time
The Times
United Press International
United States Nuclear Regulatory Commission
The Wall Street Journal
Washington Post
Washington Times

INDEX

Abortions, 43-5
Abrahamsson, K-A., 101
Academy of Sciences (Ukrainian SSR), 243, 271-2
Academy of Sciences (USSR), 85, 134, 148, 166, 272
Accidents, 113, 241-2, 259-60
Afanasyev, V.G., 134-5
Afghanistan, 178, 191
Agriculture, 60, 145-6; impact of Ch. on European, 76-8; in special zone, 78-87, 89-90, 218, 284n; in Ukraine, 272
Akhtyamov, N.G., 110
Alekseev, E., 134
Alymov, A., 210
Amosov, N.M., 209-10
Amur Oblast, 97
Andropov, Yu., 123
Antonovna, O., 201
Argentina, 6
Arkhipov, N., 79-81
Armenian nuclear power plant, 262-3
Armenian SSR, Armenians, 97, 225, 233, 262-3

Artek camp, 127, 143
Atomic Energy of Canada Limited (AECL), 16, 18
Atommash factory, 96, 101-2, 105-6
Austria, 239, 257-8
Avikson, T., 141, 183-6, 188, 190-1
Azerbaidzhan SSR, Azerbaidzhanis, 94, 225

Balakovo nuclear power plant, 106
Barringer, F., 127, 289n
Baydyuk, Y., 189-90
Bazhenov, E.O., 110-11
Belgium, Belgians, 259
Belgorod, 95
Belorussia, 45, 60, 73-4, 81-2, 84, 178, 195, 241, 266, 269
Benivka, 219
Benyukh, 118
Berdichev, 70
Berezhest River, 65
Berezovsky hydroelectric station, 92
Bespalov, K., 176
Bhopal tragedy, 259, 261

303

3   54   63   65

measures
65, 66

73
media — less in
for local peop